Previously Developed Land

Industrial Activities and Contamination

Second Edition

Paul Syms

Editorial offices:
Blackwell Publishing Ltd, 9600 Garsington Road, Oxford OX4 2DQ, UK
Tel: +44 (0)1865 776868
Blackwell Publishing Inc., 350 Main Street, Malden, MA 02148-5020, USA
Tel: +1 781 388 8250
Blackwell Publishing Asia Pty Ltd, 550 Swanston Street, Carlton, Victoria 3053, Australia
Tel: +61 (0)3 8359 1011

Published under the title *Desk Reference Guide to Potentially Contaminative Land Uses* by ISVA 1999
Second edition published 2004 by Blackwell Publishing Ltd

Library of Congress Cataloging-in-Publication Data
Syms, Paul M.
Previously developed land : industrial activities and contamination / Paul Syms. – 2nd ed.
p. cm.
Enl. ed. of: Desk reference guide to potentially contaminative land uses.
Includes bibliographical references (p.) and index.
ISBN 1-4051-0697-2 (pbk. : alk. paper)
1. Brownfields–Great Britain. 2. Brownfields–Law and legislation–Great Britain.
3. Reclamation of land–Great Britain. 4. Soil pollution–Risk assessment–Great Britain.
I. Syms, Paul M. Desk reference guide to potentially contaminative land uses. II. Title.

HD598.S95 2004
333.77–dc22

2004007272

ISBN 1-4051-0697-2

A catalogue record for this title is available from the British Library

Set in 10/13pt Palatino
by Graphicraft Limited, Hong Kong
Printed and bound in Great Britain
by T J International, Cornwall

The publisher's policy is to use permanent paper from mills that operate a sustainable forestry policy, and which has been manufactured from pulp processed using acid-free and elementary chlorine-free practices. Furthermore, the publisher ensures that the text paper and cover board used have met acceptable environmental accreditation standards.

For further information on Blackwell Publishing, visit our website:
www.thatconstructionsite.com

Contents

Foreword

Living in Hong Kong as I do, I am only too aware of the value of land as a scarce resource and, in the context of any aspiring city, of the need to find ways of releasing its latent development potential and ensuring its overall contribution to the urban fabric. This is particularly so in the case of previously developed or brownfield sites, which for centuries may have been occupied by industrial or manufacturing operations which have now fled to more suitable and cheaper environs and have left behind serious issues resulting from contamination and dereliction.

The challenges involved in bringing such land back into re-use necessitate not only skills in the areas of decontamination but also a clear understanding of all the aspects of the land conversion process, be it relating to engineering or design, land use planning, viability and financial analysis, project costing, a clear understanding of the market or development, construction, marketing and management skills. More particularly, it requires a team or consortium approach by the client advisory team and an appreciation of the roles of each and all of the professionals who can contribute to the success of such a project.

In his book *Previously Developed Land*, Paul Syms aims to provide those professionals not directly involved with remediation or treatment of contamination with a better understanding and appreciation of the role that such specialists can play and the contribution that they make. It is therefore a welcome addition to the sources of reference which support the case for a multi-disciplinary approach in such instances, bringing together and blending the best from a whole range of disciplines. In particular Paul highlights the benefits that can be achieved when addressing the redevelopment of sites, hitherto occupied by industry. Specifically, Paul is to be congratulated on presenting the case in a clear and well argued way and I am sure his book will not only be read by those wishing better to appreciate and as a result avoid the problems that might be encountered, but also by those remediation professionals who themselves require a quick reference guide.

Nicholas Brooke
President 2003–04
Royal Institution of Chartered Surveyors

Preface

This work may be seen as a substantially enlarged and revised successor to the *Desk Reference Guide to Potentially Contaminative Land Uses*, which was based on research undertaken over a six-year period up to 1999 and is now sold out. My objective has been to produce a book that will be of benefit to practitioners and students across the wide range of disciplines concerned with our physical environment. The book responds to UK Government policies that focus upon the re-use of previously developed land in preference to the use of greenfields. It is intended for use by developers undertaking, or contemplating, projects involving land re-use, as well as the members of their professional teams. Valuation surveyors should find it of benefit when deciding upon the extent to which current or previous industrial activities may have affected land values. I hope that engineers and environmental consultants will also find it to be a useful reference source.

As part of the *Desk Reference Guide* I produced a hierarchy table 'intended to represent the likelihood of encountering contaminants which will require some form of remediation'. In the preface to the guide I acknowledged that some readers 'may question, or even challenge, the ranking or risk categorisation applied to the different industrial uses' and I welcomed comments or suggestions as to how the work might be improved.

Critics of the guide were extremely small in number and their criticisms related only to questioning the assumptions used in preparing the hierarchy table. For the most part the comments received were highly appreciative, acknowledging the value of having, in a single volume, such a 'ready reference' to the potential problems that may result from our industrial legacy. A number of local authorities subsequently went on to use the guide as the means for prioritising inspections of land required as part of their strategies in implementing Part IIA of the Environmental Protection Act 1990. Comments such as these have encouraged me to produce the present work.

Previously Developed Land: Industrial Activities and Contamination builds on the earlier work by including several chapters on topics directly related to the re-use and valuation of land that has been previously developed. As with the *Desk Reference Guide* the work is divided into two parts. Part A – Issues Influencing Redevelopment and Value – opens with an introductory chapter that sets the policy context and leads into chapters on valuation, property development, recording land condition, legal issues, town planning and geographical information systems.

Several authors who are leading authorities in their fields have contributed chapters in this section. Judith Lowe, an engineer and advisor to various government departments and agencies, explains the importance of maintaining accurate records of land condition. Paul Sheridan, the Head of Environmental Law at CMS Cameron McKenna, looks at the development of environmental legislation concerning land contamination and associated matters. Ted Kitchen, Professor of Town Planning and Urban Regeneration at Sheffield Hallam University, discusses impending changes to the planning system. James Cadoux-Hudson and Donna Lyndsay, from Landmark Information Group, describe the role of Geographical Information Systems (GIS) in recording and assessing land contamination. Dr Simon Johnson, of Certa, kindly brought me up to date with regard to the availability of insurance to cover land contamination and associated risks. I should like to thank them all for their contributions.

Part B – Industrial Activities and Contamination – examines industrial activities, grouped into 39 categories, with brief descriptions of the processes involved and the contaminants that might be found on land previously used for these activities. The potential for contamination to occur exists at all stages in manufacturing activities, from the delivery of raw materials, through the manufacturing process itself to the disposal of wastes, and I should like to thank Iain McBurnie, one of my dissertation students on the BSc Environmental Management at Sheffield Hallam University for his excellent graphic, which clearly sets out the problems.

As an addition to the brief descriptions, several leading site investigation and soil remediation experts have contributed their own thoughts on particular problems that might arise, including a number of mini case studies. In particular I should like to thank Dr Tom Henman at Enviros, Mike Smith at BAE Systems Property and Environmental Services and Mike Summerskill at SEnSE Associates LLP for their contributions.

Julia Burden and Emma Moss at Blackwell Publishing have had the task of keeping me on track with the work and I greatly appreciate their support. As always, I could not have managed this undertaking without the unstinting support of my wife Janice, who made several valuable suggestions as to the form the book should take and read numerous drafts of the text.

I hope that this work will be of considerable benefit to its intended readership: environmental health officers and environmental surveyors, town planners, valuation surveyors, property developers, planning and development surveyors, environmental lawyers and everyone else with an interest in real estate and the environment. Once again I welcome comments and suggestions for improvement.

Paul Syms
Project Director
National Brownfield Strategy
English Partnerships

Biographies

Paul Syms is Project Director, National Brownfield Strategy, at English Partnerships, where he is responsible for developing and implementing the strategy to address the problems associated with England's legacy of old industrial sites. At the end of 2002 an estimated 66 000 hectares of previously developed land (PDL) in England was unused or potentially available for redevelopment; of this some 2160 sites with a total area of 16 910 hectares were identified as 'hardcore' sites, having been derelict, vacant and under-utilised for nine or more years.

Before taking up his appointment at English Partnerships in autumn 2004 Paul was Professor of Urban Land Use in the School of Environment and Development at Sheffield Hallam University. He also has extensive practical experience in the field of re-using previously developed land and buildings. Having originally qualified as a valuer, Paul subsequently obtained a Masters degree in Economic Geography from the University of Manchester and a Doctorate from Sheffield Hallam University for his research into the redevelopment and valuation of contaminated land.

Contributing Authors

James Cadoux-Hudson has been involved in the delivery of mapping and GIS services for nearly 30 years in both the UK and other countries. For the last 10 years James was involved in setting up and running Landmark Information Group Ltd, a leading provider of environmental reports that provides site-specific land and property information. The reports are generated from a large database of environmental, mapping (current and historic) and planning information, which James has been closely involved in creating.

Ted Kitchen is a Chartered Town Planner and Professor of Town Planning and Urban Regeneration at Sheffield Hallam University. Prior to joining the university in 1995, he was Director of Planning and Environmental Health at Manchester City Council. For a time he also acted as the City's Chief Executive and as Acting Clerk to the Greater Manchester Passenger Transport Authority. From 1992 to 1995 he was a non-executive director of the Greater Manchester Passenger Transport Executive. He is a member of the Best Practice Committee of the BURA and of the Priority 5 Driver Partnership, South Yorkshire Objective 1 Programme.

Judith Lowe is an expert in management of land contamination and brownfield regeneration issues. She is the co-author of a number of reports and guidance documents, including 'A Standard Land Condition Record', on behalf of the Urban Task Force working group; the Manual for the Management of Land Contamination for the Welsh Development Agency, and, most recently, 'Model Procedures' for the EA, DEFRA and others. For the last three years Judith has chaired the professional and technical panel setting multi-disciplinary standards for registration of specialists in land condition (SiLC) – an initiative sparked by the Urban Task Force.

Donna Lyndsay has an MSc in GIS with remote sensing from Greenwich University and has worked in the GIS industry for over 12 years. She currently works for Landmark Information Group as their Data Strategy Manager. In this capacity Donna manages all relationships with Landmark's Data Suppliers and sources new data for inclusion into Landmark's products.

Paul Sheridan is the partner in charge of the Environment Law Group of CMS Cameron McKenna which is universally recognised as one of the leading environment law practices. He is recognised in the directories of Chambers and Legal 500 as a leader in Environment Law. He has advised numerous UK and multi-national clients on all aspects of environment law in both contentious and non-contentious matters. He has advised on many national and trans-national corporate and property transactions, land remediations, projects, PFI/PPPs, joint ventures and privatisations and has run several environment law test cases.

Part A

Issues Influencing Redevelopment and Value

Chapter 1
Introduction and Policy Context

1.1 Introduction

> 'Previously-developed land is that which is or was occupied by a permanent structure (excluding agricultural or forestry buildings), and associated fixed surface infrastructure. The definition covers the curtilage of the development. Previously-developed land may occur in both built-up and rural settings. The definition includes defence buildings and land used for mineral extraction and waste disposal where provision for restoration has not been made through development control procedures.' (Planning Policy Guidance Note 3: *Housing* – Annex C, DETR, 2000a)

This definition excludes land and buildings that are currently in use for agricultural or forestry purposes, as well as land in built-up areas which has not been developed previously. Such excluded urban land includes parks, recreation grounds and allotments, even though these may contain buildings and other structures. Also excluded is land that was previously developed but where the remains of any structure or activity have blended into the landscape in the process of time and has thus come to be regarded as part of the natural surroundings. The amenity value of such land or the establishment of rare or endangered flora and fauna within otherwise derelict or vacant areas may outweigh arguments in favour of the re-use or redevelopment of the land.

Apart from such considerations, current Government policies in the United Kingdom strongly favour the re-use or redevelopment of land, in preference to a proliferation of greenfield developments. These policies also extend to the conversion and re-use of existing buildings for residential purposes, to the point that the demolition of a single dwellinghouse and its replacement with six new dwellings on the same site might be regarded as meeting re-use policies in respect of previously developed land.

This book is the natural successor to the earlier publication *Desk Reference Guide to Potentially Contaminative Land Uses* (Syms, 1999), albeit somewhat expanded. As with this earlier work, *Previously Developed Land: Industrial Activities and Contamination* is aimed at a readership comprising property developers, chartered planning and development surveyors, chartered environmental surveyors, chartered valuation surveyors, chartered town planners and chartered environmental health officers. It is also hoped that it will be of interest to architects, landscape architects, engineers and other professionals involved in the re-use of land. Environmental consultants and soil remediation specialists may find it of interest, not only for the 'ready reference' of industrial activities and contaminants in Part B but also for the chapters on valuation, legal issues and town planning that would normally be outside, but may have a direct effect upon, their professional activities.

The *Desk Reference Guide* was divided into two parts, with Part A considering the need for site investigations when land is to be valued or redeveloped and Part B containing brief descriptions of 39 industrial activity groups identified as most likely to have resulted in soil contamination. A similar approach has been adopted for the present work but Part A has been expanded to include chapters by leading authorities in the fields of environmental law, the recording of land condition, town planning and geographical information systems. Part B contains an introductory chapter that outlines some of the ways in which land may become contaminated through industrial activities and again includes brief descriptions of the 39 industrial activity groups. This part has also been expanded, through the addition of short case studies drawn from the experiences of the author and several highly experienced site investigation and remediation specialists.

In the *Desk Reference Guide* the author included a 'hierarchy table', which ranked the 39 groups of industrial activities according to an 'Index of Perceived Risk', reproduced in Fig. 1.1. The Index was based on research undertaken between 1994 and 1998 (see Syms, 1997a, b) and was 'intended to represent the potential for contaminative substances to be present at concentrations which will require remedial action to be taken if the site is to be redeveloped' (Syms, 1999, p. 11). In the preface to the *Desk Reference Guide* the author acknowledged that some people 'may question, or even challenge, the ranking or risk categorisation applied to the different industrial uses' but, in spite of this, the hierarchy table was welcomed by a number of readers. A confirmation

	HAZARD RANK	LAND USE CLASSIFICATION	INDEX OF PERCEIVED RISK	PERCEIVED RISK CATEGORY
CLASS A	1	Asbestos manufacture and use	1.00	HIGH
	2	Organic and inorganic chemicals production not included elsewhere	0.93	HIGH
	3	Radioactive materials processing and disposal	0.88	HIGH
	4	Gasworks, coke works, coal carbonisation and similar sites	0.85	HIGH
	5	Waste disposal sites, including hazardous wastes, landfills, incinerators, sanitary depots, drum and tank cleaning, solvent recovery	0.85	HIGH
	6	Oil refining, petrochemicals production and storage	0.84	HIGH
	7	Manufacture of pesticides	0.83	HIGH
	8	Pharmaceutical industries, including cosmetics and toiletries	0.82	HIGH
	9	Fine chemicals, dyestuffs and pigments manufacturing	0.82	HIGH
CLASS B	10	Paint, varnishes and ink manufacture	0.79	HIGH
	11	Animal slaughtering and by-products, including soap, candle and bone works; detergent manufacture	0.78	HIGH
	12	Tanning and leatherworks	0.77	HIGH
	13	Metal smelting and refining, including furnaces and forges, electro-plating, galvanising and anodising	0.74	HIGH
	14	Explosives industry, including fireworks manufacture	0.73	HIGH
	15	Iron and steelworks	0.72	HIGH
	16	Scrap yards	0.68	HIGH
	17	Engineering (heavy and general)	0.66	MEDIUM
CLASS C	18	Rubber products and processing	0.65	MEDIUM
	19	Tar, bitumen, linoleum, vinyl and asphalt works	0.65	MEDIUM
	20	Concrete, ceramics, cement and plaster works	0.65	MEDIUM
	21	Mining and extractive industries	0.65	MEDIUM
	22	Electricity generating (excluding nuclear power stations)	0.64	MEDIUM
	23	Film and photographic processing	0.63	MEDIUM
	24	Manufacture of disinfectants	0.62	MEDIUM
	25	Paper and printing works, including newsprint (usually excludes 'high street' printers)	0.60	MEDIUM
	26	Glass manufacture	0.58	MEDIUM
	27	Fertiliser manufacture	0.58	MEDIUM
	28	Timber treatment works	0.58	MEDIUM
	29	Sewage treatment works	0.54	MEDIUM
	30	Garages, inc. sale of automotive fuel, repair of cars and bikes	0.53	MEDIUM
	31	Transport depots, road haulage, commercial vehicle fuelling, local authority yards and depots	0.53	MEDIUM
	32	Railway land, including yards and tracks	0.53	MEDIUM
	33	Electrical and electronics manufacture, inc. semiconductor manufacturing plants	0.48	MEDIUM
	34	Textiles manufacture and dyeing	0.48	MEDIUM
	35	Laundries and dry-cleaning (larger scale, not usually 'high street')	0.48	MEDIUM
CLASS D	36	Plastic products manufacture, moulding and extrusion; building materials; fibre glass, fibre glass resins and products	0.48	MEDIUM
	37	Dockyards and wharves	0.48	MEDIUM
	38	Food processing, including brewing and malting, distilling of spirits	0.45	LOW
	39	Airports and similar	0.45	LOW

Fig. 1.1 Hierarchy table of industrial activities. Reproduced from *Desk Reference Guide to Potentially Contaminative Land Uses.*

of its usefulness was its adoption by several local authorities as the basis for their strategies for inspections to be carried out under Part IIA of the Environmental Protection Act 1990.

The only real substantive criticisms made of the hierarchy table were that it was not based on scientific facts, which is acknowledged as it was based on experiences and perceptions, and that it was not capable of taking into account when, and over what duration, the industrial activities were carried out, as well as other site-specific factors such as the underlying geology and the vulnerability of ground and surface waters. It is almost always the case that the longer ago the industrial activities took place, the greater will be the likelihood of contamination being present when compared to more modern operations of similar types and scale. This is because environmental awareness has increased significantly since the 1980s and many waste disposal practices have also changed over the same period. It is also the case that the more historic contamination may, according to its type, have undergone a greater degree of degradation than contaminants that have entered the ground more recently and may therefore be reduced in concentrations or toxicity. The need for each property to be assessed on a site-specific basis was stressed in the *Desk Reference Guide*, advice that is reiterated here.

The research that underpinned the 'Index of Perceived Risk' was based on asking site investigation and remediation experts about those activities that were, in their opinion, most likely to present problems and to require remediation prior to redevelopment. In addition to being ranked in the Index the sites were also classified into four groups:

- Class A: intrusive investigation strongly recommended
- Class B: intrusive investigation recommended
- Class C: intrusive investigation desirable
- Class D: intrusive investigation optional

Since the *Desk Reference Guide* was published, the emphasis on the redevelopment of previously developed land has increased and organisations such as the National House-Building Council now require geotechnical and geoenvironmental investigations to be carried out on both brownfield and greenfield land if insurance cover is to be provided for new developments. As a result of these changes, intending developers and their advisers should no longer regard intrusive investigations (including the laboratory analysis of soil and water samples) as being either desirable or optional – instead they should be seen as being essential. It is not the purpose of the present work to consider site investigations, site assessment, risk analysis and remediation/treatment options. For a discussion of these aspects of re-using previously developed land (PDL), intending developers, their surveyors and other advisers are recommended to read Chapters 5, 6 and 7 of *Land, Development and Design* (Syms, 2002).

The purpose of the rest of Chapter 1 is to provide the reader with an outline of present government policies relating to the re-use of PDL in preference to greenfield development. It includes references to the roles of English Partnerships, the English Regional Development Agencies (RDAs), Welsh Development Agency, Scottish Enterprise, the National Land Use Database (NLUD), the Commission for Architecture in the Built Environment (CABE), and to other organisations. The significance of Part IIA of the Environmental Protection Act 1990 (Part IIA) is discussed and an outline of the Contaminated Land Exposure Assessment (CLEA) guidance is also provided.

Valuers of land and buildings are often faced with the problem of having to decide upon the likely impact that the presence of contaminative substances may have on the value of premises that they are instructed to value. Very often valuers are expected to provide opinions as to value without having the benefit of a site investigation report upon which to base those opinions. It is usually impractical, in terms of both time and cost, for the valuer to decline to provide a valuation, or to defer issuing the report, until such time as a full site investigation report is available. Valuation issues, including international research and practices, as well as current guidance are considered in Chapter 2.

Previously developed land covers a wide range of sites used for myriad purposes. The focus of this book is on those sites that have been used for industrial activities and these, for ease of reference, have been grouped under the same 39 headings as used in the *Desk Reference Guide* but the reader should appreciate that these encompass several hundred different industry types and processes. Very often, industrial activities are regarded as the worst, or only, sources of contamination. That is not necessarily the case and it should be remembered that by no means all industrial sites are contaminated, some are and have been managed to very high environmental standards, and not all contamination originates from industrial activities. Other land uses, such as hospitals and laboratories, may also result in contamination. For example, several years ago the author was involved in advising several householders whose new homes were found to have been constructed on a hospital waste tip, containing medicine bottles, syringes and dental equipment, requiring replacement of the soils in the back gardens to depths of between three and four metres. Readers should also be aware that there may be many other barriers or obstacles that need to be overcome before land can be re-used or redeveloped; these are discussed in Chapter 3.

The legislation relating to the identification and remediation of contaminated land is described briefly later in this chapter. Land so seriously affected by the presence of contamination that it falls to be determined as 'contaminated land' or as a 'special site' under this legislation represents only a very small percentage of previously developed, or 'brownfield', land that is available for re-use or redevelopment, as illustrated in Fig. 1.2. Nevertheless, other previously developed land, whether derelict, vacant or under-used, may be affected by different legislation, either domestic or European, and the laws relating to land use are continually changing. In Chapter 5 Paul Sheridan, the Head of Environment Law at CMS Cameron McKenna, examines the legal issues surrounding the re-use of land.

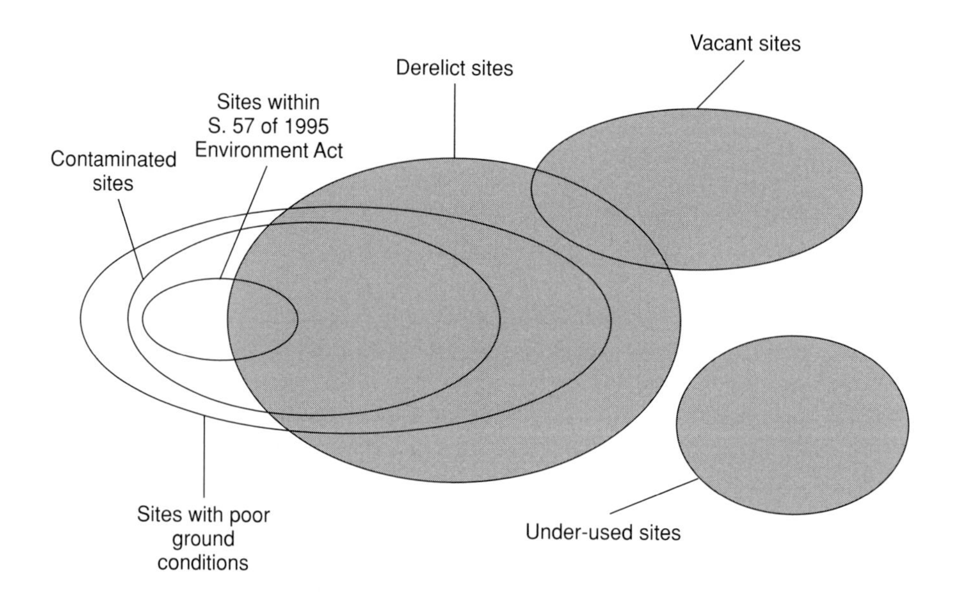

n.b. not to scale: the area of the figures is not proportional to the land area.
Brownfield land is that shown by the shaded areas.

Fig. 1.2 Although not intended to be to scale, this diagram from the Parliamentary Office of Science and Technology report *A Brown and Pleasant Land* places land that is 'contaminated' in the legal sense in context with other issues that relate to previously developed land. Reproduced with permission from Post: *A Brown and Pleasant Land*, report 117.

The Urban Task Force in its final report, *Towards an Urban Renaissance*, recommended establishing:

> 'a national framework for identifying, managing and communicating the risks that arise throughout the assessment, treatment and aftercare of contaminated and previously contaminated sites.' (Urban Task Force, 1999, recommendation 75, p. 242)

The Task Force then went on to suggest that once this framework was in place, 'there would be a strong case for securing even greater consistency in the handling of site information by introducing a form of standardised documentation.' It also recommended the piloting of:

> 'standardised Land Condition Statements, to provide more certainty and consistency in the management and sale of contaminated and previously contaminated sites.' (Urban Task Force, 1999, recommendation 76, p. 243)

In order to take forward these recommendations and, in particular, to agree a standardised format for the recording of information, a working party was set up under the joint Chairmanship of Phil Kirby, then Environment Director of BG Properties (now SecondSite Property, see www.secondsite-property.com) and Lorna Walker, a director of Arup, both of whom had been members of the Urban Task Force. This working party brought together about 20 organisations representing different interests in land and property development. The Technical Convenor of the working party was Judith Lowe, an independent consultant and former head of the Contaminated Land Branch at the Department of the Environment, who in Chapter 4 discusses the importance of information relating to ground conditions.

Town planning in Britain, in its 'modern' form owes its origins to the Town and Country Planning Act 1947. This has seen many revisions, as well as successor Acts of Parliament, and the Government has decided that a thorough overhaul of the system is required. The Green Paper *Planning: Delivering a Fundamental Change* (DTLR, 2001) was summarised in Syms (2002, Chapter 16), and in Chapter 6 of the present work Ted Kitchen, Professor of Town Planning and Urban Regeneration at Sheffield Hallam University and former Chief Planning Officer of Manchester City Council, discusses how the changes are being implemented.

Geographical Information Systems (GIS) provide invaluable tools in enabling prospective developers and their teams to pinpoint, with a high degree of accuracy, historical contamination and other potential obstacles to redevelopment. The use of digital data with GIS has significantly simplified the process of compiling, collating and analysing information relating to land use. However, as with all computer-based systems, the output is only as good as the information that has been input in the first place. In Chapter 7, James Cadoux-Hudson and Donna Lyndsay, both from Landmark Information Group (see www.landmarkinfo.co.uk) provide the reader with an introduction to GIS and explain its usefulness in the assessment of previously developed land.

1.2 Overview of government policy on sustainable development and previously developed land

Previously developed land offers significant opportunities and it is the Government's policy objective to ensure that these opportunities for new developments are exploited in preference to extending development further into the countryside (see Fig. 1.3). In adopting such a policy the Government 'recognise that previously developed land is

Fig. 1.3 Government policies favour the re-use of previously developed land for housing, as in the case of these new houses on the site of a former engineering works, but densities are often little different to those for greenfield development.

not always available in rural areas facing genuine development needs, and this can lead to hard choices being made' (ODPM, 2003a, para 4.9). With this one caveat therefore, policy is firmly focused on the re-use and redevelopment of previously developed land.

In 2001, 61% of all new housing was built on previously developed land[1], i.e. exceeding the target of 60% set in Planning Policy Guidance Note 23 (PPG23) in 2000, but the Government is concerned that, where greenfield land has been needed for housing, it has not always been used efficiently. Currently the average density for new build on greenfield land in the South East is 22 dwellings per hectare. If all future development planned for the South East were built at 30 dwellings per hectare, this would save an area the size of Peterborough over the next 15 years (ODPM, 2003a, background). Therefore, it is a Government objective to 'ensure land is not used in a profligate way' (ODPM, 2003a, para 4.8). In order to address this issue, planning applications to develop larger sites for new housing at below 30 dwellings per hectare in areas of high demand will be liable to be called in and will have to be justified following public enquiry.

The target that 60% of additional homes should be on previously developed land is to be maintained, although some would argue that this should be increased. The Campaign to Protect Rural England in May 2003 expressed the view that 'by sticking to its 60% target, the Government risks slowing further progress and sacrificing countryside to unnecessary greenfield development' (CPRE, 2003). In the view of the CPRE the target should be increased, aiming for at least 75% of new housing on previously developed land.

Through the Regional Development Agencies and English Partnerships (see below), the Government aims to remediate brownfield land at a rate of over 1400 hectares per year for economic, commercial, residential and leisure use. This is an area the size of the London Borough of Islington. In addition to ensuring the reuse of previously developed land, the Government wishes to see more development activity being focused on the revitalisation of city and town centres. Besides previously developed land being redeveloped for housing use, in 2000 new in-town shopping floorspace exceeded out-of-town development for the first time since the early 1990s. The Government also will ensure that every local authority has undertaken an urban capacity study (as set out in PPG3) to identify the full potential for using previously developed land and conversions (ODPM, 2003a, section 4).

[1] The 2001 estimate was subsequently increased from the provisional figure of 61% published in LUCS-17 (ODPM, 2002c) to 63% published in LUCS-18 (ODPM, 2003b). Analysis showed that the provisional 2002 estimate was unlikely to be revised outside of the range 63–67%.

Although the policy emphasis is on using previously developed land to meet the requirements for new developments, the Government recognises that not all such land is capable of being developed. English Partnerships, Groundwork, the Forestry Commission and the Environment Agency will work together to create the Land Restoration Trust to restore and manage brownfield land that is suitable only for use as public green space. The Trust will work in partnership with local communities.

1.3 Development agencies

Intending developers of previously developed land in the United Kingdom can turn to regional development agencies for support and advice in identifying suitable sites for development. The support provided by these agencies will differ from region to region but may include advice on appropriate remediation methods, help with site acquisition and access to public funds to support remediation projects. In addition to the regional agencies mentioned below, there are a number of Urban Regeneration Companies and local development agencies that may be able to provide support for development projects and web addresses for a number of these are provided at the end of the book.

1.3.1 English Partnerships

English Partnerships, the national regeneration agency, is the Government's expert advisor on previously developed land and is a key delivery agency for the urban renaissance and the Government's Sustainable Communities agenda. Supporting high quality sustainable growth across England, English Partnerships is a major force in ensuring that full and effective use is made of previously developed land to secure the urban renaissance (see www.englishpartnerships.co.uk).

England is one of the most densely populated countries in the world so it is particularly important to make the best possible use of the scarce supply of land. One of the most efficient ways of achieving this is by focusing on the re-use of previously developed land (brownfield land). In 2002, English Partnerships, at the request of the Government, was asked to develop and maintain a National Brownfield Strategy (see English Partnerships, 2003), which in addition to providing detailed information on the supply of brownfield land in England, will establish a co-ordinated approach on how to allocate regeneration resources. The National Brownfield Strategy will aim to identify key national projects for public intervention and provide a coherent vision for the future development of brownfield land in a way that underpins national, regional and local development aspirations. The strategy will also form a key element in helping the Government meet its Public Sector Agreement targets for building new homes on brownfield land, thereby reducing the pressure on greenbelt.

A prerequisite for developing the National Brownfield Strategy is high quality information on the availability and development potential of brownfield land. The National Land Use Database (NLUD), developed by English Partnerships, the Office of the Deputy Prime Minister, the Improvement and Development Agency and Ordnance Survey, is the principal source of intelligence on the current stock of previously developed land. The NLUD Database Report for 2002 produced an estimate of 'previously developed land that is unused or may be available for redevelopment' in England of 66 100 hectares, of which 29 000 hectares were potentially available for housing (Tables 1.1 and 1.2). Of this, some 17 000 hectares has lain derelict for nearly a decade (NLUD, 2003).

According to local authority estimates, the previously developed land identified in 2002 as being potentially available for housing could accommodate 880 000 new dwellings. The report also stated that about 9% of previously developed land (PDL) in 2001 had been redeveloped in 2002 (NLUD, 2003). The National Brownfield Strategy will build on the work of NLUD in presenting accurate and up-to-date information on the distribution and the development capacity of brownfield land.

A further measure being developed by English Partnerships to ensure best use is made of brownfield land is the creation and maintenance of a Register of Surplus Public Sector Land. Announced by the Government in the Sustainable Communities Plan, the Register will require all government departments to record their surplus sites on the Register prior to marketing the sites for sale to give other public sector bodies, including English Partnerships, an opportunity to identify new uses for the land in accordance with local development strategies. This Register will be shared across Government and in the long term may have the ability to be more widely accessible.

Whilst the emphasis is in supporting housing and regeneration strategies, not all brownfield sites are suitable for development. Working with the Environment Agency, Groundwork and the Forestry Commission, English Partnerships is developing a new venture, the Land Restoration Trust, to tackle enduring dereliction across England and transforming eyesores into valuable new 'green amenities' such

Table 1.1 Previously developed land[1] by land type, planned use and suitability for housing: England 2002. Reproduced with permission from ODPM Statistical Release: Previously Developed Land that may be available for development (brownfield sites) in 2002.

Land/building type	Planning allocation or permission[2] (hectares)				**All previously developed land that is unused or may be available for redevelopment**	of which suitable for housing[3] hectares/numbers	
	Housing	Mixed use	Other	None		Total area	Number of dwellings[4]
Vacant and derelict land and buildings							
Previously developed vacant land	2 530	2 260	6 570	4 330	15 680	5 800	180 500
Derelict land/buildings	1 530	2 490	8 860	7 070	19 960	5 680	149 400
Vacant buildings	1 600	510	1 090	1 880	5 070	2 770	112 200
All vacant and derelict land and buildings	**5 660**	**5 260**	**16 510**	**13 270**	**40 710**	**14 250**	**442 080**
Currently in use							
Allocated in a local plan for any use or with planning permission for any use	5 490	5 140	5 950	—	16 570	9 430	272 400
Known redevelopment potential but no planning allocation or permission	—	—	—	8 830	8 830	4 840	169 800
All currently in use	**5 490**	**5 140**	**5 950**	**8 830**	**25 400**	**14 280**	**442 160**
All land types	**11 150**	**10 400**	**22 460**	**22 100**	**66 110**	**28 520**	**884 200**

[1] Previously developed land that is unused or may be available for development. Excludes previously developed land within the National Parks.
[2] Land that has planning permission or is allocated in a local plan for the use indicated (includes some allocations in draft plans).
[3] Based on sites judged by the local authorities to be suitable for residential development.
[4] Based on existing planning permissions or estimated capacity based on current density assumptions.
Note: Rounding errors as in orginal.

Table 1.2 Previously developed land[1] by land type and Government Office Region: England 2002. Reproduced with permission from ODPM Statistical Release: Previously Developed Land that may be available for development (brownfield sites) in 2002.

	Vacant and derelict land			Currently in use hectares		**hectares**
Government office region	Previously developed vacant land	Derelict land and buildings	Vacant buildings	Allocated in a local plan for any use or with planning permission for any use	Known redevelopment potential but no planning allocation or permission	**All previously developed land that is unused or may be available for redevelopment**
North East	1 950	1 630	230	560	420	**4 780**
North West	2 630	5 610	1 070	1 490	970	**11 770**
Yorkshire & the Humber	2 080	3 270	840	820	980	**8 000**
East Midlands	1 160	2 470	730	1 300	730	**6 390**
West Midlands	2 250	1 710	420	1 390	780	**6 560**
East of England	1 350	1 740	490	2 050	1 910	**7 540**
London	390	460	280	1 920	460	**3 520**
South East	2 300	1 430	560	5 320	1 300	**10 910**
South West	1 560	1 630	460	1 720	1 280	**6 650**
England	**15 680**	**19 960**	**5 070**	**16 570**	**8 830**	**66 110**

[1] Previously developed land that is unused or may be available for development. Excludes previously developed land within the National Parks.

Note: Rounding errors as in original.

as woodland, parks, commons, nature areas and other public open spaces providing benefits for both people and nature.

1.3.2 *Regional Development Agencies in England*

Deputy Prime Minister John Prescott announced the new programme for the regions in December 1997, launching the Regions White Paper, *Building Partnerships for Prosperity* (DETR, 1997). The Regional Development Agencies (RDAs) were established under the Regional Development Agencies Act 1998 and the first eight came into being on 1 April 1999 (the ninth, London, followed later in July 2000). The Government's objective in establishing the nine RDAs was to bring a new focus to economic development and to build prosperity across the English regions. The RDAs are non-department public bodies

(NDPBs) with a primary role as strategic drivers of regional economic development. They aim to co-ordinate regional economic development and regeneration, enable the English regions to improve their relative competitiveness and reduce the imbalance that exists within and between regions. Each RDA has five statutory purposes, which are:

- to further economic development and regeneration
- to promote business efficiency, investment and competitiveness
- to promote employment
- to enhance development and application of skills relevant to employment
- to contribute to sustainable development

The agenda of the RDAs includes regional regeneration, taking forward regional competitiveness, taking the lead on regional inward investment and, working with regional partners, ensuring the development of a regional skills action plan to ensure that skills training matches the needs of the labour market (DTI website, October 2003 – www.consumers.gov.uk/rda/info). The percentage of new homes built on previously developed land, as a measure of sustainable development, is one of nine 'core indicators' against which the performance of all RDAs is judged. The net area (in hectares) of derelict land brought back into use is an 'Activity Indicator' for the RDAs, as a measure of the physical regeneration achieved.

The RDAs will develop English Partnership's National Brownfield Strategy in more detail to produce Brownfield Land Action Plans, in co-operation with local authorities, and other relevant agencies and statutory bodies. These plans will fit closely with the Regional Economic Strategies and Regional Housing Strategies (ODPM, 2003a, para 4.6).

1.3.3 *Welsh Development Agency*

The Welsh Development Agency is sponsored by the Welsh Assembly Government and is the leading enabler of business support in Wales. The agency works for the people of Wales by stimulating more competitive businesses in vibrant communities. (Welsh Development Agency website, October 2003 – www.wda.co.uk).

The agency is committed to creating an environment in which industry and Wales can flourish by providing a range of commercial and industrial property services, including:

- Key Sites and premises database
- Capital Projects
- Property Services

Availability of the right kind of land, in the right locations, is the first essential ingredient of profitable and beneficial property development. Identifying land suitable for all forms of development including sites for employment, housing, leisure and town centre schemes, the WDA Land Development and Legal Services team can exercise powers of compulsory purchase and title cleansing to support the growth of prosperity in Wales.

1.3.4 *Scottish Enterprise*

Rated among the world's top economic development agencies, Scottish Enterprise is the main economic development agency for Scotland, covering 93% of the population from Grampian to the Border (Scottish Enterprise website, October 2003 – www.scottish-enterprise.com).

Along with twelve local enterprise companies, Scottish Enterprise works in partnership with the private and public sectors, with the aim of securing the long-term future of the Scottish economy by making its industries more competitive. The aims of Scottish Enterprise are to:

- help businesses start up and assist existing companies to grow
- promote and encourage exporting
- attract inward investment
- develop skills

The priorities of Scottish Enterprise include commercialisation of academic ideas into good business opportunities, e-business, globalisation and economic inclusion. Scottish Enterprise provides free, independent, advice through its local enterprise companies on finding, buying, leasing, building and even developing commercial property. A list of available properties is also maintained on the agency's website.

1.3.5 *Invest Northern Ireland*

Northern Ireland is a small region with unique educational, cultural and environmental strengths. Invest Northern Ireland aims to harness those strengths by encouraging innovation and entrepreneurship to create an environment in which companies – whether home grown or from overseas – will flourish (Invest Northern Ireland website, October 2003 – www.investni.com).

The agency believes that by stimulating sustainable demand for new employment and services, new levels of wealth will be created for the benefit of all who live in the region. It aims to encourage existing businesses to widen their horizons, develop external markets and to embrace the ethos of business improvement. The agency also offers a warm welcome to globally-orientated overseas companies.

1.4 Other organisations

In addition to the development agencies referred to in the previous section, there are a number of other non-government organisations that may be able to assist intending developers contemplating the redevelopment of previously developed land. These include the Commission for Architecture in the Built Environment and Contaminated Land: Applications in Real Environments.

1.4.1 *Commission for Architecture in the Built Environment*

Although possibly having a less direct involvement in seeking the re-use of previously developed land, the Commission for Architecture in the Built Environment (CABE) may well have an important role to play in how developers go about the redevelopment exercise. CABE is an executive non-departmental public body, funded by the Department for Culture, Media and Sport (DCMS). The Commission believe that good architecture, landscape architecture, urban design and spatial planning can have a positive impact, enriching our culture, raising our spirits and making life more fun. It can breathe life back into places suffering from economic and social decline, restore community identity and civic pride as well as attracting investors and visitors. In the view of CABE, the delivery of better public services can be achieved through a quality environment, thereby having a direct impact on the quality of healthcare, education, welfare and other services.

Furthermore, the Commission believe that value for money in design can be promoted by reducing the lifetime costs of buildings and improving their performance. Good design can help to reduce crime and anti-social behaviour by creating places that foster community ownership and eliminate physical opportunities for vandalism, violence and theft. Taken together these beliefs and values may be seen as seeking to promote more environmentally friendly ways of living. The Commission seek to achieve their vision by (www.cabe.org.uk):

- 'Campaigning until every child is being educated in a well-designed school and every patient treated in a decent healthcare environment.
- Fighting alongside the public for greater care and attention in the design and management of our parks, playgrounds, streets and squares.
- Helping people who are looking to buy or rent a new home by improving the design quality of new houses and neighbourhoods.
- Building up the evidence that good design creates economic and social value, so that investment in good design is seen as a necessity, not a luxury.
- Keeping every client of a new building on their toes by demanding design quality in projects of all shapes and sizes in all parts of the country.
- Working across the country with all those involved in the planning, design, construction and management process so that opportunities for design innovation are always exploited.
- Thinking ahead to the new demands that will be made of our built environment in 10 or 20 years time, through drivers such as climate change, technological advance and demographic change.'

1.4.2 *Contaminated Land: Applications In Real Environments (CL:AIRE)*

CL:AIRE was formed specifically to find innovative practical solutions to deal with the disparate issues relating to land contamination and is a public/private partnership involving the following stakeholders:

- government policy makers
- regulators
- industry
- research organisations
- technology developers

It provides a link between the main players in contaminated land remediation in the UK, to catalyse the development of cost-effective methods of investigating and remediating contaminated land in a sustainable way (CL:AIRE website, October 2003 – www.claire.co.uk/aboutframes.html).

One of the objectives of CL:AIRE is to develop a network of contaminated sites throughout the UK for the purpose of demonstrating remedial technologies. With this objective in mind, CL:AIRE is interested in all sites that are representative of industrially contaminated land in the UK and is interested to hear from landowners or others involved with contaminated land, and who may be willing to allow a research project or technology demonstration to be carried out on it. There is no limit on the size of the site but CL:AIRE prefers that the site be characterised and that the type and level of contamination of the soil and groundwater be known prior to the approach being made.

Project duration on a site will vary, with some research projects requiring sites for up to three years to allow for long term monitoring; however, technology vendors may only require a site for a few weeks or months to carry out a demonstration. CL:AIRE recognises that some sites may be sensitive to publicity and, therefore, site locations can be kept confidential. CL:AIRE is interested only in understanding the type of ground conditions and the nature and extent of contamination.

A number of sites that have been put forward to CL:AIRE are brownfield sites that include plans to carry out full-scale remediation as part of redevelopment. CL:AIRE should be involved at the early stages of these developments to develop an appropriate work plan (CL:AIRE website October 2003 – www.claire.co.uk/aboutframes.html).

1.5 Environmental Protection Act 1990, Part IIA

Government policy in tackling the problems associated with land contamination is that, wherever possible, these should be dealt with as part of the planning and development process of re-using or redeveloping previously developed land. Nevertheless it recognises that this is not possible to achieve in all circumstances and has therefore introduced legislation to deal with the worst cases of land contamination, although even then it seeks to encourage 'voluntary' action to remediate land so as to make it suitable for use or to overcome the harmful effects of contamination.

Otherwise known as the 'contaminated land legislation', Part IIA of the Environmental Protection Act 1990 came into force in England in April 2000 by way of section 57 of the Environment Act 1995. Local authority environmental health departments are the primary regulators under the legislation and the local authorities were given fifteen months to prepare their strategies for implementation. Under the legislation all local authorities have a duty to inspect their areas and to identify land that may be contaminated in accordance with the following definition (Environmental Protection Act 1990, s.78A(2)):

'any land which appears to the local authority in whose area it is situated to be in such a condition, by reason of substances in, on or under the land, that –

(a) significant harm is being caused or there is a significant possibility of such harm being caused; or
(b) [significant][2] pollution of controlled waters is being, or is likely to be, caused'

In the absence of voluntary action by the polluter or site owner, local authorities will be required to serve remediation notices on anyone who caused or knowingly permitted the presence of the contaminants. This could result in non-polluting owners or occupiers being liable for the cost of remediating the site to a standard appropriate to its existing use and at which no significant environmental harm is being, or is likely to be, caused.

If contamination migrates, other persons such as neighbouring landowners could sue the person responsible, particularly in respect of a financial loss, such as a property sale not proceeding. In respect of those sites designated as 'special sites' the Environment Agency is the lead enforcement authority. Alternatively the Agency could clean up the pollution then sue for the recovery of costs.

To be classified as 'contaminated', it is necessary for the regulatory authority to identify at least one *significant pollutant linkage*, whereby a *contaminant* can travel, or be transmitted via a *pathway* to a *receptor* (target). The subsequent regulations and guidance deal with four receptor groups: human beings, protected ecological environments, domestic animals and crops, and buildings. The extent to which 'harm' may be deemed 'significant' is given in statutory guidance issued by the Department of the Environment, Transport and the Regions – see www.environment.detr.gov.uk.

1.5.1 *Who is responsible for the cost of dealing with contamination?*

Having determined that a contamination problem exists, the regulator must then identify an 'appropriate person' to bear the remediation cost, using the 'polluter pays' principle. In cases where there is more than one appropriate person, six exclusion tests have to be applied, as listed below.

[2] Clause 79 of the Water Act 2003 amended the definition of contaminated land by inserting the word 'significant' before 'pollution of controlled waters'.

- *Test 1: Excluded Activities.* The purpose of this test is to exclude from liability those people and/or organisations that have had a connection with the land but have not been directly responsible for the polluting operations. This would include providing (or withholding) financial assistance through making a loan or grant; underwriting an insurance policy; consigning waste to another person who assumes responsibility for its disposal but then disposes of it in an improper manner; creating a tenancy or licence over the land in question; issuing any statutory permission, taking (or not taking) enforcement action in respect of the land in question; and providing legal, financial, engineering, scientific or technical advice, including an intrusive site investigation (provided that the investigation itself is not the cause of the contamination event in the significant pollution linkage).
- *Test 2: Payments Made for Remediation.* The purpose of this test is to exclude from liability those who have, in effect, met their responsibilities by making payment to another member of the liability group, and where the payment would have been sufficient at the date it was made to pay for the remediation in question but the remediation was not carried out, or was not done effectively.
- *Test 3: Sold With Information.* The purpose of this test is to exclude from liability those who, although they have caused or knowingly permitted the presence of a significant pollutant, have disposed of the land in question in circumstances where it is reasonable that another member of the liability group should assume responsibility for the remediation. The transaction must be at arm's length, with no residual interest being retained by the vendor, and the buyer must have been provided with sufficient information to make them aware of the presence on the land of the pollutant identified in the significant pollutant linkage in question.
- *Test 4: Changes to Substances.* The purpose of this test is to exclude from liability persons who caused or permitted the presence in, on or under the land, of a substance which has only led to the creation of a significant pollutant linkage because of its interaction with another substance which was later introduced to the land by another person.
- *Test 5: Escaped Substances.* The purpose of this test is to exclude from liability those who would otherwise be liable for the remediation of contaminated land which has become contaminated as a result of the escape of substances from other land, where it can be shown that another member of the liability group was actually responsible for that escape.

- *Test 6: Introduction of Pathways or Receptors*. The purpose of this test is to exclude from liability those who would otherwise be liable solely because of the introduction by others of the relevant pathways or receptors in the significant pollutant linkage. (*Source*: DETR Circular 02/2000, pp. 111–119)

The parties are divided into two liability classes: Class A, comprising all possible polluters, and Class B, the landowners. Once the tests have been completed, the regulator must then apportion the costs appropriately.

With support from the Department for Food, Environment and Rural Affairs (DEFRA) and the Environment Agency, a local authority 'procedural' guide to applying the Contaminated Land regime has been produced together with training for local authority staff. The work was led by a steering group, which included representatives from DEFRA, the Local Government Association, the Chartered Institution of Environmental Health and the Environment Agency. The guide is only available via the CIEH website, at www.cieh.org.uk/research/environment/epalaguide.htm. Although the guide is intended for use by local authorities, it has the added benefit of providing developers and their consultants with information as to how local authorities should proceed if a potential development site is suspected of falling within the scope of the legislation.

A developer who acquires a site but does not continue the polluting activity would normally be a Class B person. However, if engineering activities result in the creation of a pathway between the contaminants and receptor, say ground water, then this results in the developer being classed as a Class A person. Also, if the developer knowingly permits the polluting activity to continue this could result in inclusion in Class A. These factors will need to be borne in mind by the valuer as the liability as a Class A or Class B person will differ greatly. Not only will these potential liabilities affect a developer's decision as to whether or not to proceed with the redevelopment of a contaminated site but they will also affect the perception of valuation professionals.

In addition, a landlord would need to ensure that tenants do not cause contamination of either land or groundwater. A landlord would also need to ensure that suitable indemnification is in place against any clean-up liabilities caused by a tenant. Surveyors acting as property managers will therefore need to be very aware of the activities of tenants. Similarly, banks, insurance companies, building societies and any other financial institutions will need to be aware that they may have potential liabilities under this legislation.

1.6 Contaminated Land Exposure Assessment (CLEA)

The United Kingdom was one of the first countries to introduce, in the early 1980s, the use of 'trigger concentrations' for contaminants in soil, which depended upon the intended use of the site to assist in determining the significance of contamination (ICRCL, 1987, p. 1). If the trigger concentrations were exceeded this would prompt further investigation and may potentially lead to decisions being taken as to the need for remedial action. The Interdepartmental Committee on the Redevelopment of Contaminated Land, in its paper *Guidance on the Assessment and Redevelopment of Contaminated Land* (ICRCL 59/83), issued values for 'spot' samples based on an adequate site investigation carried out prior to development in respect of 19 contaminants, plus soil acidity (ICRCL, 1987, p. 17). These included contaminants that were potentially harmful to human health, those harmful to plant life and others that may have deleterious effects on buildings or their services.

The second edition of the ICRCL guidance, published in 1987, updated the earlier guidance and also replaced four earlier reports, which were withdrawn. The committee also published a number of other papers, dealing with specific types of contamination or land uses, such as metalliferous mining, asbestos, scrap yards, sewage works, gasworks and landfills. Over a period of almost two decades the trigger concentrations in ICRCL 59/83 were widely used as part of the decision making process in respect of potentially contaminated development sites. They were often widely misused by consultants, developers and regulating authorities, all of whom sometimes failed to adequately relate the trigger values to specific site conditions, or to proposed end uses, deciding upon the need for remediation without proper consideration of exposure pathways. The range of contaminants covered by the guidance was also limited and site investigators sometimes fell into the trap of commissioning laboratory analyses against the 'standard ICRCL' suite, without giving adequate thought to the possibility that current or former activities on a site may have resulted in the presence of contaminants not covered by the guidance.

Therefore, whilst the usefulness of the ICRCL guidance as a decision making tool was widely recognised, its limitations were also acknowledged. In 1992, in response to a report by the House of Commons Select Committee on the Environment (House of Commons, 1990), the then Department of the Environment initiated research to develop a scientific framework for assessing the risks to human health from land contamination (Environment Agency, 2003a). This research resulted in the development of the Contaminated Land Exposure Assessment (CLEA) model, which together with the related toxicology reviews and the first Soil Guideline Values (SGVs), was launched in March 2002. The CLEA model estimates child and adult exposures to soil contaminants for those potentially living, working and or playing on contaminated sites. It assumes long-term exposure to the contaminants.

In December of the same year ICRCL guidance note 59/83 was withdrawn as being 'technically out of date' and 'not in line with the current statutory regime (Part IIA of the Environmental Protection Act 1990) and associated policy' (DEFRA, 2002). The other ICRCL guidance documents were not withdrawn as it was 'recognised that many of the other notes provide more general information useful to the assessor, relevant to contaminants not considered by the CLEA package' (DEFRA, 2002) and the Department hopes to review these with a view to up-dating and re-presenting their contents as appropriate. Thus any site investigations and remediation decisions taken after December 2002 should be based on the Soil Guideline Values developed for use with the CLEA model, or upon similar toxicologically based data where SGVs have not yet been published, instead of being reliant upon ICRCL 59/83.

The 'CLEA package' comprises four Contaminated Land Reports (CLRs 7 to 10), together with Soil Guideline Values and related Tox papers (see Box 1.1). As at December 2003, Tox and SGV papers were in preparation, or proposed, for a further 38 contaminants. Readers are also recommended to refer to the earlier Contaminated Land Reports (CLRs 1 to 6) (Box 1.2) for authoritative and researched advice on how best to identify and assess the problems contamination can pose and what can be done to tackle them. It must, however, be stressed that every site is unique and that all site-specific conditions need to be considered in relation to published guidance.

It is not the purpose of this chapter to provide a detailed technical review or critique of the CLEA package, rather to present an overview that will be of benefit to readers who are not soil scientists but who may have to deal with land contamination issues in their capacities as architects, developers, town planners and surveyors. The following sections therefore contain brief summaries of the four newer Contaminated Land Reports.

Box 1.1 CLEA publications.

CLR7: Assessment of risks to human health from land contamination: an overview of the development of Soil Guideline Values and related research introduces the other new reports in this series. It describes the legal framework (including the statutory definition of contaminated land under Part IIA of the Environmental Protection Act 1990), the development and use of Soil Guideline Values; and references to related research. DEFRA/EA. March 2002.

CLR8: Priority contaminants for the assessment of land identifies priority contaminants (or families of contaminants). These have been selected on the basis that they are likely to be present at many sites affected by current or former industrial use in the United Kingdom in sufficient concentrations to cause harm; and that they pose a risk either to humans, buildings, water sources or ecosystems. The report also indicates which contaminants are associated with particular industries. DEFRA/EA. March 2002.

CLR9: Contaminants in soils: collation of toxicological data and intake values for humans sets out the approach used to derive tolerable daily soil intakes and index doses for contaminants, which were subsequently used to determine Soil Guideline Values. DEFRA/EA. March 2002.

CLR10: Contaminated Land Exposure Assessment Model (CLEA): technical basis and algorithms describes the conceptual exposure models for each standard land use for which Soil Guideline Values are derived. It sets out the technical basis for modelling exposure and provides a comprehensive reference to all the default parameters and algorithms used. DEFRA/EA. March 2002.

'Tox' series: The first series of toxicological reports detail the derivation of tolerable daily soil intakes or index doses for the first set of contaminants for which Soil Guideline Values have been determined.

Collation of Toxicological Data and Intake Values for Humans [name of contaminant]

TOX1 *Arsenic*. DEFRA/EA. March 2002.
TOX2 *Benzo(a)pyrene*. April 2002.
TOX3 *Cadmium*. March 2002.
TOX4 *Chromium*. March 2002.
TOX7 *Mercury*. March 2002.
TOX8 *Nickel*. March 2002.
TOX9 *Phenol*. October 2003.
TOX10 *Selenium*. March 2002.
TOX11 *Benzene*. April 2003.

continued

Box 1.1 (*continued*)

TOX5 *Inorganic cyanide.* March 2002.
TOX6 *Lead.* March 2002.
TOX12 *Dioxins, furans and dioxin-like PCBs.* March 2003.

'SGV' series: The first series of the Soil Guideline Values reports sets out the derivation of Soil Guideline Values for the first set of contaminants for which toxicological data have been determined:

Soil Guideline Values for [name of contaminant] contamination in soils

SGV1 *Arsenic.* DEFRA/EA. March 2002.
SGV3 *Cadmium.* March 2002.
SGV4 *Chromium.* March 2002.
SGV5 *Inorganic mercury.* March 2002.
SGV7 *Nickel.* March 2002.
SGV9 *Selenium.* March 2002.
SGV10 *Lead.* March 2002.

Source: Contaminated Land Report 7 and the DEFRA website.

Box 1.2 Contaminated Land Reports 1–6.

CLR 1 **A framework for assessing the impact of contaminated land on groundwater and surface water.** Report by Aspinwall & Co. Volumes 1 & 2. DoE, 1994.

CLR 2 **Guidance on preliminary site inspection of contaminated land.** Report by Applied Environmental Research Centre Ltd. Volume 1. DoE, 1994. Volume 2. DoE, 1994.

CLR 3 **Documentary research on industrial sites.** Report by RPS Group plc. DoE, 1994.

CLR 4 **Sampling strategies for contaminated land.** Report by The Centre for Research into the Built Environment, The Nottingham Trent University. DoE, 1994.

CLR 5 **Information systems for land contamination.** Report by Meta Generics Ltd. DoE, 1994.

CLR 6 **Prioritisation & categorisation procedure for sites which may be contaminated.** Report by M J Carter Associates. DoE, 1995.

1.6.1 CLR 7: Assessment of risks to human health from land contamination

This report provides an overview of the development of Soil Guideline Values and related research. It sets out (Environment Agency, 2002a, pp. 1–2):

- The legislative context for the use of CLRs including the Part IIA contaminated land regime and the planning system.
- The use of Soil Guideline Values in the assessment of chronic risks to human health arising from land contamination.
- Approaches to soil sampling, testing and data analysis when using Soil Guideline Values in different regulatory contexts.
- A summary of the next three CLRs, which describe in detail the Soil Guideline Values and the technical basis for their derivation, as well as other information relevant to assessment of risks to human health.

Although the information contained in this and the other reports, together with the SGVs, will be of assistance to individuals and regulatory authorities seeking to determine whether or not land is contaminated in accordance with the definition of 'contaminated land' in Part IIA, it is not intended solely for that purpose. The information relates primarily to the assessment of land contamination on human health over an individual's lifetime but it also recognises that short-term exposures, such as in the case of transient risks to building site workers, are also a cause for concern.

1.6.2 CLR 8: Priority contaminants for the assessment of land

This report provides a selection of contaminants that may be relevant for the assessment of contaminated land. The contaminants are relevant because they are likely to be found on a large number of industrial sites in the UK and have the potential to affect human health and the environment (Environment Agency, 2002b, p. 1). The selected contaminants will be of interest to developers and their advisers, as well as to regulators, because the contaminants concerned are deemed to be important on a national basis. They can also be used as part of the site assessment process although, it must be noted, investigation of all the selected substances will not be necessary for every site and equally not all possible contaminants for every industrial use are included in the report.

Two criteria were used for the selection of potential contaminants:

- The contaminants must be likely to be present on many sites in current or former industrial use in sufficient concentrations to cause harm; and
- The contaminants must pose a potential risk to human beings and to sensitive environmental receptors.

Consequently, the substances selected are (Environment Agency, 2002b, pp. 3–4):

- 'Likely to occur on many sites in sufficient concentrations to cause harm or pollution; and
- Known or suspected to pose a significant risk of harm to humans (death, serious injury, cancer or other disease, genetic mutation, birth defects, or the impairment of reproductive functions); or
- Known or suspected to pose a significant risk in the water environment, or likely to cause other adverse effects in the water environment, as a result of their presence on land; or
- Known or suspected to pose a significant risk to ecosystems as a result of their presence on land; or
- Known or suspected to have significant effects on buildings or building materials; or
- Known or suspected to be persistent and mobile in soils or to have a tendency to bioaccumulate through exposure of sensitive organisms.'

The substances found to occur frequently on sites have been assessed to establish whether they pose an actual risk. Risk will depend on the nature of the hazard, the probability of exposure, the pathway by which exposure occurs, and the likely effects on the receptor. Each substance has therefore been assessed to establish whether it is likely to be a hazard and whether there is a potential pathway to each of the receptors mentioned previously (Environment Agency, 2002b, p. 4). Tables 1.3 and 1.4, respectively, set out the potential inorganic and organic contaminants for the assessment of industrial land and the receptors that might be affected.

1.6.3 CLR 9: Contaminants in soil: collation of toxicological data and intake values for humans

This report describes a framework for the collation of toxicological data to support the derivation of soil contaminant intakes that are

Table 1.3 Inorganic contaminants for the assessment of industrial land.

	Contaminants			
Receptors at risk	Metals	Semi-metals and non-metals	Inorganic contaminants	Others
Humans	Beryllium, Cadmium, Chromium, Lead, Mercury, Nickel, Vanadium	Arsenic, Selenium, Sulphur	Cyanides (free and complex)	Asbestos
Water	As for humans, plus Barium, Copper and Zinc	Arsenic, Boron, Selenium	Cyanides, Nitrate, Sulphate, Sulphide	—
Vegetation/ ecosystem	Beryllium, Cadmium, Copper, Lead, Mercury, Nickel, Zinc	Boron, Selenium, Sulphur	Cyanides, Sulphate, Sulphide	—
Construction materials	None	Sulphur	Complex Cyanide, Sulphate, Sulphide	—

Notes: This should not be regarded as a comprehensive list; actual site conditions and past or present uses, including the alkalinity/acidity of the soil should always be taken into account.
Source: based on Table 2.1, Contaminated Land Report 8 (CLR8), Environment Agency, 2002b.

protective of human health. It involved assessing the potential harm to human health from contaminants in soil and using this information for deriving *health criteria*. Tolerable daily intakes (TDIs) and minimal risk levels (termed 'Index Doses') are derived for threshold and non-threshold[3] contaminants (Environment Agency, 2002c, p. 1). The report was written for technical professionals who are familiar with the risks posed by land contamination but who may not necessarily be experts in toxicology.

[3] Chemicals for which a threshold for health effects cannot be assumed, such as genotoxic carcinogens and mutagens.

Table 1.4 Organic contaminants for the assessment of industrial land.

Receptors at risk	**Contaminants**				
	Aromatic Hydrocarbons	Chlorinated Aliphatic Hydrocarbons	Chlorinated Aromatic Hydrocarbons	Organmetallics	Other
Humans	Benzene, Chlorophenols, Ethylbenzene, Phenol, Toluene, *o*-Xylene, *m,p*-Xylene, Polycyclic Aromatic Hydrocarbons (PAHs)	Chloroform, Carbon Tetrachloride, Vinyl chloride, 1,2-Dichloroethane, 1,1,1-Trichloroethane, Trichloroethene, Tetrachloroethene, Hexachlorobut-1,3-diene Dieldrin	Chlorobenzenes, Chlorotoluenes, Pentacholophenol, Polychlorinated biphenyls, Dioxins and Furans	Organolead compounds, Organotin compounds	Acetone, Oil and fuel hydrocarbons
Water	As humans	As humans	As humans	Organotin compounds	As humans
Vegetation/ ecosystem	As humans, except for PAHs	As humans, except for vinyl chloride	As humans	Organolead compounds	—
Construction materials	As for vegetation and ecosystem	Carbon Tetrachloride, 1,2-Dichloroethane, 1,1,1-Trichloroethane, Trichloroethene, Tetrachloroethene	—	—	—

Notes: This should not be regarded as a comprehensive list; actual site conditions and past or present uses, including the alkalinity/acidity of the soil should always be taken into account.
Source: based on Table 2.2, Contaminated Land Report 8 (CLR8), Environment Agency, 2002b.

1.6.4 *CLR 10: The Contaminated Land Exposure Assessment (CLEA) model: technical basis and algorithms*

This describes the technical principles of the CLEA model. The report contains explanatory material that is important for use of the Soil Guideline Values. In particular it sets out the underlying assumptions that have been made to predict exposure for three standard land-use scenarios: residential, allotments and commercial/industrial (see Box 1.3). It aims to assist risk assessors to judge whether it is appropriate to use a Soil Guideline Value as part of the assessment for a particular site, taking account of land-use contaminant type and general site conditions (Environment Agency, 2002d, p. 1).

Box 1.3 Standard land uses for Soil Guideline Values.

Residential
People live in a wide variety of dwellings including, for example, detached, semi-detached and terraced property up to two storeys high. This takes into account different house designs including detached, semi-detached and terraced buildings up to two storeys in height, with either suspended floors and ground-bearing slabs. The land use assumes that residents have private gardens and/or access to community open space close to the home. Exposure to contaminants has been estimated both with and without the eating of homegrown vegetables. These two alternatives represent the key difference in potential exposure to contamination between people living in a house with a garden, where vegetables might be grown, and those living in a house without a private garden. The conceptual model for exposure to contaminants arising from residential sites assumes that the site consists of a single property within a square-shaped area and that water for drinking and bathing is obtained from the mains supply. It assumes that exposure pathways will include ingestion of soil and of household dust, dermal contacts with soil and household dust, inhalation of fugitive soil dust and household dust, inhalation of vapour both inside and outside the dwelling. For dwellings with private gardens, the pathways are extended to include the ingestion of contaminated vegetables and of contaminated soil attached to vegetables. The critical receptor is assumed to be a young female child with the duration of exposure covering the first six years of life.

continued

Box 1.3 *(continued)*

Allotments
Allotments occupy more than 10 000 hectares in England and total nearly 300 000 in number, used by more than a quarter of a million people and their families. This land use assumes that allotments, commonly made available by the local authority or through member owned associations, will be used by local people for the purpose of growing fruit and vegetables for their own consumption. Some allotment holders also keep animals including rabbits, hens and ducks but potential exposure to contaminated meat and eggs has not been considered for this land use. The conceptual model for exposure to contaminants from allotments assumes that the number of visits made will vary according to age and season, but with working adults making an average two visits per week, with a worst case duration of four hours per visit. The site is assumed to be largely open space with individual plots and, for the standard allotment scenario, as with residential land use the critical receptor is assumed to be a young female child with the duration of exposure covering the first six years of life.

Commercial/industrial
This land use assumes that the workplace is a permanent single-storey building, where employees spend most time indoors, whilst recognising that the modern economy has a diverse range of industrial and commercial enterprises. The commercial/industrial site is assumed to occupy a total square-shaped area with the office or factory occupying 0.1 hectare of this space. It assumes that the employees are involved in office-based or relatively light physical work. This land use is not intended to include sites involving 100% hard cover, such as car parks, where the risks to the site-user from ingestion or skin contact are minimal while the hard surface remains intact. For the purpose of the conceptual model the exposure pathways are assumed to include ingestion of soil and of building dust, dermal contacts with soil and building dust, inhalation of fugitive soil dust and building dust, inhalation of vapour both inside and outside the building. In terms of exposure, it is assumed employees work a five-day week, on average 38.1 hours, and have up to six weeks leave (including statutory holidays), with the critical receptor being a female worker.

Source: based on sections 3 and 4 Contaminated Land Report 10 (CLR 10) Environment Agency, 2002. Reproduced with permission from The Environment Agency.

1.6.5 CLR 11: Model procedures for the management of land contamination

This report has yet to be published by the Environment Agency, although a consultation draft has been issued. The Model Procedures:

> 'are intended to provide the framework for a structured technical process for informing decisions about land contamination. The basic technical process can be adapted to apply in a range of regulatory and management contexts, subject to the specific constraints set by these contexts.
> The Model Procedures are intended to assist all those involved in "managing" the land, in particular landowners, developers, industry, professional advisers, financial service providers, planners and regulators.' (Environment Agency, 2003b, p. 6)

1.7 Summary

Government policies are firmly focused on giving priority to the reuse of previously developed land, especially for housing, and to a lesser extent on strengthening the centres of towns and cities. Some emphasis is being placed on increasing the density of residential developments, with developers and local authorities being faced with the possibility that schemes might be called in by the Deputy Prime Minister if the development density is less than 30 units per hectare. Whilst the possibility of applications being called in does raise the possibility of development projects being delayed, with attendant financial implications, the density level involved is hardly a radical departure from the average level already being achieved and does little to improve the efficient use of land.

The long awaited CLEA guidance is now in use, albeit with Soil Guideline Values and toxicity data in respect of a very limited suite of contaminant, and the ICRCL guidance has been withdrawn. Further values and guidance have yet to be published and it will be some time before CLEA is fully operational.

Checklist

- Is the property in question to be redeveloped or is the existing use expected to continue?
- If the site is to be redeveloped, does the proposed development accord with current land-use policies?
- Do the contaminants analysed as part of the site investigation relate to current and/or previous uses on the site?
- Was the site investigation report prepared after December 2002 and, if so, has the CLEA model or a similar approach been used?
- Consult with Regional Development Agencies for support and advice.
- Would information on the site assist CL:AIRE in compiling their database?
- Is there a likelihood that contamination may have migrated to/from adjoining sites?
- Is the site likely to be determined as 'contaminated land' or as a 'special site' under Part IIA of the Environmental Protection Act 1990 and, if so, has an 'appropriate person' been identified to bear the remediation costs?
- Would the landowner or developer be a Class A or a Class B person?

Chapter 2

Approaches to Valuation

2.1 Introduction

Whilst Chapter 1 focused mainly on the redevelopment aspects of previously developed land and provided the policy context, this chapter examines issues and practicalities surrounding the valuation of land that may have suffered harm as the result of its current or historic usage. By no means all such land will fall within the scope of the Part IIA legislation described in Chapter 1 and, indeed, much of the land in question may be suitable for its existing use, not causing harm to workers, visitors to the premises or to the wider environment. However, unless the valuer is provided with appropriate environmental reports and is trained to understand their meanings and implications, how is he or she going to be able to assess the extent to which current or historic land uses may impact upon value?

Parts of this chapter are based on research commissioned by the RICS Foundation. The research was commissioned with the intention of enabling practitioners to understand some of the implications involved in the valuation of land affected by the presence of contamination. The full report (Syms & Weber, 2003), which includes comparisons with practice in the United States and Australasia, is available from the RICS Foundation.

As mentioned in Chapter 1, Government policy with regard to the remediation and/or treatment of land contamination is that, wherever possible, it should be dealt with as part of the planning and development process. In other words, unless contaminants in the ground are actually causing significant harm, or the pollution of controlled waters (or there is a strong possibility that significant harm or pollution will occur), then no action needs to be taken until such time as the site comes to be redeveloped. It is also the case that the extent of remediation and the cost needed to make an industrial site safe for its existing use, under Part IIA, may be considerably less than that required to make the same site suitable for redevelopment. This has important implications for the valuers of industrial and commercial premises, as the cost of voluntary compliance with the Part IIA legislation so as to ensure the continued safe use of the site for industrial purposes may be well below that of remediating the same site in order to construct a new factory. To give an example, for the ongoing use of the site and its existing buildings it may be sufficient to prevent the contaminants from coming into contact with site workers or visitors by simply covering the affected soil with concrete, or even an engineered clay cap. When considering the construction of a new factory, or even an extension to the existing buildings, the health and safety of site workers and new substructures and services will have to be taken into account. Building Control, the local authority's environmental health officer, the Environment Agency and the Health and Safety Executive may all become involved. Therefore, it may be necessary to embark upon a programme of remediation, which may include, for example, bio-remediation or soil vapour extraction of the contaminants whilst still in the ground, excavation and ex-situ treatment, such as soil washing, off-site disposal or re-interment within the site in a secure 'coffin'.

From the valuer's perspective it may be a relatively simple exercise to obtain cost estimates for the works needed to comply with the legal requirements under Part IIA. But it may be a totally different matter to reflect in the valuation the limitations imposed on the use of the site, even for continued industrial purposes, which may arise out of the need to undertake remediation or treatment works if the buildings are to be extended or replaced. Should the valuer be expected to make assumptions as to ways in which buildings may be extended or replaced over years to come, or in respect of remediation or treatment techniques that might be employed? The answer to these questions must surely be no, as the making of such assumptions would normally be outside the scope of the valuer's remit and competence. Yet, failing to reflect the limitations on use may have a very great impact on value, potentially turning a positive value that assumes continuation of existing activities without any building works into a negative value if the cost of decontamination is taken into account. All of this assumes continuation of industrial activities on the site and does not even start to consider the (potentially much higher) cost of preparing the site for a more sensitive use, such as housing with private gardens in which vegetables might be grown.

The two main sources of information for valuers in the United Kingdom are the guidance notes and practice statements within the

RICS Appraisal and Valuation Standards (the Red Book) (RICS, 2003a) and the RICS Environment Faculty Guidance Note 'Contamination and Environmental Matters' (RICS, 2003b).

It is essential for anyone who undertakes valuations of property, land or machinery to have a copy of the Red Book. The book contains all official RICS practice statements and guidance notes on valuation, which are mandatory on all RICS members. Most banks and other property lending institutions require valuations to be undertaken by chartered surveyors, or at least to comply with RICS guidance. Use of the Red Book ensures that valuers follow best practice and are compliant with the law. So far as 'Contamination and hazardous substances' are concerned, the Red Book states that 'a valuer will not normally be competent to advise on either the nature, or risks, of contamination or hazardous substances, or any costs involved with their removal'. The Red Book does however qualify this statement by adding 'where valuers have prior knowledge of the locality and experience of the type of property being valued, they can reasonably be expected to comment on the potential that may exist for contamination, and the impact this could have on value and marketability' (RICS, 2003a, Appendix 2.2).

'Contamination and Environmental Matters' provides guidance to surveyors across a range of environmental matters, including legal issues, the identification of contamination and the valuation of land affected by contamination. The guidance note makes the very important point that 'when a surveyor is approached to provide a valuation service, he or she must consider whether contamination, or the potential for land to be contaminated are factors that need to be taken into account' (RICS, 2003b, p. 39). It is therefore incumbent upon the valuer to consider the possibility that contamination might have an impact on value and, if the valuer is not going to take contamination issues into account, this must be clearly stated in the valuation report. The valuer will also need to consider the possibility that remediation may be required at some future date, assuming that the land is suitable for its existing or proposed use. If this is likely to be the case, the valuer will need to report upon the contingent liabilities that may be involved and consider the need for a sinking fund to meet these future costs.

Although the RICS Red Book and Contamination guidance note provide a considerable amount of information that is useful to the valuer, for example identifying factors that should be taken into account in the valuation, neither document gives any real guidance as to the valuation methods that might be employed in taking into account the effects of contamination or other environmental aspects of previously developed land. The rest of this chapter considers the different approaches to valuation that have been championed by academics and practitioners.

2.2 Recent and current research

A number of very different valuation approaches have been considered by researchers and practitioners, including the sales comparison approach (Patchin, 1994), the cost approach (Wilson, 1994), and the cash flow approach (Wilson, 1996). Some of these, for example the sales comparison approach and the cost approach are most suited to valuations required for owner occupation or redevelopment purposes, rather than for investment properties. Several other researchers have considered the impact of contamination on investment properties, for example work by Lizieri *et al.* (1995), Richards (1996) and Kennedy (1998). Whilst these valuation methods may be appropriate for investment properties they are not necessarily suitable for valuing contaminated land held as part of the assets of manufacturing companies. The valuation of manufacturing assets, and their subsequent redevelopment, was considered by Syms (1997a). As the reader will therefore grasp, the choice of a method appropriate to the type of property/interest to be valued is most important.

Some of the methods concentrate on the effect contamination may have on investment yields, whereas others consider the costs associated with tackling the contamination, whether the expenditure has to be incurred immediately or deferred to the future, and any possible 'stigma' effect on the value. Most authors have considered the question of 'stigma', in respect of which Wiltshaw (1998) observed that there is no universal consensus as to its cause.

2.2.1 *Campanella*

The first author to discuss valuation techniques for contaminated properties was Campanella (1984) writing in the US, who noted that the sales comparison approach often became the appraiser's most convincing indication of value in an ordinary, i.e. non-contaminated, valuation assignment. He also noted that it is very rare to find objective and truly comparable sales of contaminated property. Campanella therefore considered the income capitalisation approach to be limited in its reliability, as the valuer would have to make sure that the comparable has the same type, quantity and location of pollution to add up to the same threat as the subject.

According to Campanella, the principle of substitution leads to reliance upon the 'cost-to-cure' approach. Using this approach the loss in value is considered to equal the cost of curing the contamination problem but, as he warned, the potential for error in its use and the quantification of lost value by this method can be a highly case-specific task.

2.2.2 *Patchin*

Patchin (1988) discussed the range in contamination impact on value by noting that properties seriously affected by toxic contamination are usually unmarketable and that those less seriously affected are subject to at least reduced marketability, i.e. they are 'stigmatised'. He noted that the amount of loss in market value varied according to the nature and extent of the contamination and listed three broad categories of losses in value for contaminated properties:

(1) costs of cleanup
(2) liability to the public
(3) stigma after cleanup

The valuation method that he suggested is the mortgage–equity technique of capitalisation, where the capitalisation rates are dependent on three factors:

(1) the equity yield rate
(2) mortgage terms available
(3) anticipated future appreciation or depreciation

On this basis, according to Patchin, the capitalisation rate for contaminated properties can be adjusted according to the particular circumstances, such as:

- the extent and nature of the contamination
- whether the property is industrial, commercial or special purpose, etc.
- available financing
- the demand for alternate uses

Patchin then suggested three scenarios, which can be summarised as follows (Patchin, 1988):

> **Scenario 1:** the risk of future liabilities after remediation is such that the contamination is serious enough to render the property unmarketable, with only a 'value in use'.
>
> **Scenario 2:** this assumes that the property has limited marketability, with a major difference in value due to the nature and extent of the contamination.
>
> **Scenario 3:** the implications of contamination are such that the effect on value arises out of a change in the highest and best use of the land subject to the presence of contamination.

Patchin's basic technique was to first value the property as if not contaminated, then in its contaminated condition. He subsequently subtracted the value 'after' from the value before contamination to estimate the damages. The 'after' value was derived by increasing his capitalisation rate in the first two examples, presumably to reflect stigma, but no support was given for the rate chosen. In Scenario 3 he deducted his estimated land value for storage use from an estimate of land value for office use to arrive at his impaired value. However, this example does not include a deduction for stigma.

Patchin introduced two new concepts, which are a 'potential liability' and the possibility of 'stigma'. He stated that his solution to the problem of accounting for these concepts in the valuation is to use proper capitalisation rates. The major problem is once again the lack of data that can be used to properly utilise the income approach. All of the scenarios seem to require a high degree of accuracy in remedial cost estimates in order to inform the valuation process. In practice, this information on remedial costs may not be readily available.

2.2.3 *Wilson*

Wilson (1992) provided a good example of the need to discuss remedial cost in terms of probability, stressing the importance of knowing the most probable cost to control a problem. He defined the most probable cost as being the one considered to be most likely, which may be regarded as taking the cost estimate, 'expected cost', for the preferred treatment method (or combination of methods) and adjusting it to allow for any savings that might accrue from the cautious approach adopted by the tendering contractor. Use of this approach also requires the valuer to know the possible range of costs, from the position of achieving maximum savings, with everything going right, to the maximum additional cost if everything goes wrong. Thus an expected (or estimate cost) might be £2.8 million but the most probable cost £2.4 million, within an overall range of £2.0 million to £3.0 million. Clearly this approach might well be unacceptable from the viewpoint of an intending developer, who would prefer to see a little more certainty in terms of figures rather than such a broad range. Although possibly not ideal for the developer, a discussion of this nature may well be appropriate to a valuation report, especially if the valuer is being expected to reflect possible contamination impacts in the report with the benefit of only limited information as to the actual land condition.

Wilson also referred to the 2X factor in which buyers tend to deduct twice the estimate of 'most likely' remedial cost due to the historically proven uncertainty of these estimates. In other words, a cautious contingency provision to cover unforeseen items, such as contamination being more extensive or more toxic than originally envisaged.

2.2.4 *Mundy*

Mundy (1992a) described the total loss in value resulting from contamination as often being more than the 'cost to cure', suggesting that property affected by contamination may suffer from diminution in value from two factors. The first is real risks that can be scientifically quantified and the second is perceived risks, which can be more difficult to quantify and he too suggested that the public in the marketplace would adopt a 2X approach, doubling the expected remedial cost as part of a cautious approach.

2.2.5 *Lizieri* et al.

Lizieri *et al.* (1995) considered the relationship between Valuation Methodology and Environmental Legislation in a report for the RICS Education Trust and noted that the standard method of valuation in the United Kingdom is still based on the use of an *all risks yield*, derived from analysis of market comparables. The researchers observed that 'despite criticisms of the methodology, it remains the technique most commonly used in valuation, both for determining likely selling price and for portfolio performance and lending security purposes' (Lizieri *et al.*, 1995, p. 32). They went on to ask, 'How well does the all risks yield approach deal with environmental hazard?'

So far as investment properties, or those charged as security for bank lending, are concerned, they suggested that when considering environmental impact on a property valuation, an investor should review both the impact of potential future cash flow costs of holding the asset *and* the market valuation impacts of other investors' perceptions of environmental risk. They formed the view that it is possible for these sums to differ, particularly in a 'new' market situation, i.e. a market in contaminated properties, characterised by uncertainty and with few comparable transactions.

As with Wilson and Mundy, Lizieri and his colleagues concluded that investors would take a cautious approach, by both deducting costs and shifting yields upwards, thereby 'double counting' environmental risk, to account for stigma and unknown future liabilities.

2.2.6 *Richards*

Richards (1995) conducted an interview survey of representatives from a range of sectors perceived to be most active, or likely to have a significant input to contaminated land valuation and risk assessment. The interviews focused on a number of key areas relating to contaminated land, including:

- the issues which should be considered prior to providing valuations or other advice
- the main 'cost factors' to be taken into account in preparing a valuation
- the relative suitability and adaptability of different valuation methods
- the appropriateness of different valuation bases
- issues relating to marketing and transfer
- lending and investment issues
- redevelopment issues

He concluded that if the client required the impact of contamination to be taken into account in the valuation, the valuer should advise that an environmental consultant be employed and that only when such advice is available should a valuation be prepared. Recommendations such as this may not be acceptable to clients, especially in situations where valuations are required within a short timescale, e.g. for bank financing or in defence of a takeover bid. However, it is a fact that many valuers are neither competent to undertake valuations that fully reflect environmental conditions, nor will they carry suitable indemnity insurance. There are also significant cost implications, as the fees for a full environmental study may be many times the valuation fee.

Richards' study considered several valuation methods and a majority of interviewees indicated that they would employ a 'residual-type' valuation approach. That is the method by which the amount of money remaining to the developer, for land acquisition (and/or development profit), is calculated after deducting the cost of carrying out the development (including all financing and other expenses) from the resultant investment value or sale price. With the impact of contamination depending upon the cost of remediation this will have a significant impact on the cost of carrying out the development and may even reduce the residual sum to a negative figure.

As with a number of other commentators, Richards noted the problems associated with obtaining comparable income and yield figures, with the valuer having to decide whether to use comparable information from impaired (contaminated) or unimpaired properties. He pointed out that one danger in using information from 'impaired' properties is that it is highly unlikely that two contaminated sites will be sufficiently alike as to be truly comparable. In view of this problem, he suggested that it might be more realistic to use unimpaired comparable information and then use site-specific information in order to adjust the figures and arrive at a valuation based upon the 'cost to correct'.

Another valuation method considered by Richards and his interviewees was that of adjusting the *All Risks Yield (ARY)* and he suggested that, by including implicit assumptions relating to contamination (in addition to existing assumptions such as growth, risk of illiquidity and risk of tenant default), it may be possible to reflect both remedial costs and 'stigma'.

Although Open Market Value is the most widely used basis for commercial valuations, the valuers amongst Richards' interviewees were asked whether they thought *Estimated Realisation Price (ERP)*, could provide an appropriate basis for the valuation of contaminated land. Valuers were mainly unconvinced as to the logic behind this. *Existing Use Value* was considered to be an appropriate basis for valuing properties where the contamination was not so severe as to prevent continuation of the existing use whereas, conversely, the *Alternative Use Valuation* basis would imply a need for some remediation, resulting in a reduced value for the property.

In spite of all the problems associated with obtaining comparable information for contaminated land valuations, few of the interviewees in the Richards' study used the *Depreciated Replacement Cost (DRC)* basis of valuation. They preferred to reserve this method for the valuation of specialist industrial units, which cannot be valued by the use of comparable information.

In a second study by Richards (1997) the respondents were asked to indicate the methodologies which they would use to produce *market valuations* (MV) and *calculations of worth* (CoW), given a choice of:

- traditional (ARY) methods
- discounted cashflow (DCF) approaches
- other

The majority indicated that they would use the traditional approach to produce open market valuations and DCF approach when calculating worth.

2.2.7 *Kennedy*

Kennedy (1998) also undertook a survey of practitioners in an attempt to identify the methods used to value contaminated land. The four methods considered were:

- Direct Capital Comparison Method (DCCM)
- Investment/Income Method (IIM)
- Profits Method (PM)
- Residual Method (RM)

The most popular method was found to be IIM and the least popular valuation method was the Profits Method.

2.3 Valuation of 'non-investment' properties

It is clear from the Lizieri study, the two Richards studies and the Kennedy study, that there is no commonality of agreement between practitioners and researchers as to the appropriate method(s) to be employed in valuing investment interests in contaminated land. Even when seemingly uniform approaches are employed, as in the 'Valuation Scenarios' used by Richards, wide ranges of adjustments are employed in order to take account of the impact of contamination.

Whilst the research undertaken by Lizieri *et al.*, Richards and Kennedy focused primarily on the impact of contamination on investment properties, Syms' research (Syms, 1997a) tended more towards the valuation of contaminated assets used for manufacturing purposes and the redevelopment of previously used, and contaminated, land. The types of properties considered by Syms are rarely, if ever, found in the portfolios of major property companies and property investing institutions but are more likely to be owned by the occupier or its parent company. These include most, if not all, of the uses considered in Part B of this book.

Even though they are not normally traded as investments, properties occupied for these industrial uses have to be valued from time to time, for example for inclusion in company accounts or as security for bank borrowings. Comparable information, including rental and yield data, may be sparse or completely non-existent, therefore Direct Capital Comparison (DCCM) and Investment/Income methods are not appropriate. The 'three-stage' approach adopted by Syms (1997a, pp. 186–97) was first to prepare an *unimpaired* valuation of the land element only, disregarding any buildings, fixed plant or other structures on the site. This site valuation may be arrived at by a variety of methods including direct comparison with industrial land in the area or by deducting the Depreciated Replacement Cost of buildings etc. from the total unimpaired asset value. Having arrived at an uncontaminated land value, the method then used the 'cost to correct' approach, to arrive at an *impaired* land value, following the guidance contained in Guidance Note 2 of the RICS Appraisal and Valuation Manual (RICS, 1995), to take account of the costs involved in remediating the site. These costs included:

(a) clean-up of on-site contamination
(b) effective contamination control and management measures
(c) re-design of production facilities
(d) penalties and civil liabilities for non-compliance
(e) indemnity insurance for the future
(f) the avoidance of migration of the contamination to adjacent sites
(g) the control of migration from other sites
(h) the regular monitoring of the site

(RICS, 1995, p. 8; Syms, 1997a, p. 188)

After reducing the land value of the property by the 'cost to correct', Syms then suggested that, in the case of a valuation required for asset purposes i.e. not requiring immediate action to comply with legal requirements to make the site safe for continued use, the 'costs' could be deferred to the end of the economic life of the buildings, at which time the remediation works would be undertaken in order to facilitate redevelopment.

In the example asset valuation provided by Syms, no costs were included for items (c) process re-design and (d) penalties/civil liabilities. He argued that this is because these should be regarded as immediate costs, not capable of deferment to the end of the economic life of the buildings and all such 'non-deferrable costs should be treated as current liabilities in the valuation and specifically reported upon' (Syms, 1997a). Syms then went on to suggest that there may be a case for including part of the contamination control and management measures as 'immediate costs' and that it may be appropriate to consult with the company's auditor 'so as to determine which, if any, of the liabilities should be treated as general liabilities of the business, rather than related to the specific property'[1] (Syms, 1997a).

The impaired land value arrived at by deducting the costs required to deal with the contamination was then adjusted in the last of the three stages, to take account of possible stigma. The percentage deduction for stigma was arrived at by drawing comparisons with sales and valuation data on other contaminated sites. By this method it was not necessary to have true comparables in terms of contamination types and potential receptors, the method only required comparison with sites for which similar amounts of information were known about the severity of the contamination, with the actual stigma impact being calculated through a process of extrapolation. In a further refinement of the method, Syms (1997a, pp. 197–203) proposed a 'risk assessment' model for the third stage of the valuation, whereby the stigma impact was calculated by reference to the degree of 'risk' associated with the industrial activities that had taken place on the site.

Kennedy (1998, p. 139) criticised the 'deferment' aspect of Syms' valuation method on the basis that, even when a decision is taken to delay remedial works, some ongoing expenditure will be required to ensure contamination problems do not increase and possibly increase liabilities. Thus, in the view of Kennedy, costs of monitoring and other contamination control and management measures will be a periodic liability from the date on which contamination problems are discovered and deferment will result in their impact on value being understated. This criticism disregarded the fact that the valuation was for asset purposes of a chemical manufacturing plant and the explanation by Syms that some costs should be regarded as liabilities of the business and not direct property costs. The same applies to pre-remediation monitoring costs. The only monitoring liabilities that should be included in the valuation are those that relate directly to the remediation itself.

Bell (1998) discussed theory versus techniques, but noted that special care should be taken in the review of remediation costs, because the original cost estimates are often exceeded. In his view, the cost may require the addition of a contingency factor reflecting a complete and reasonable cost estimate, so that the real estate market may be assured that all reasonable remediation costs are accounted for in the estimates. He suggested that the contingency factor should relate to the hard costs

[1] Syms has subsequently advocated all such costs, not relating directly to the physical nature of the site, being included as 'operating' as opposed to 'real estate' liabilities.

of remediation and should not be confused with intangible losses, such as responsibility or stigma.

Bell's opinion of not including an incremental amount to cover remedial cost overruns as part of a stigma calculation differed from the opinions of previous authors. In Wilson and Mundy's calculation of stigma they included remedial cost overruns in the definition of perceived risk. Patchin (1992) provided four examples of valuation techniques that used increased equity yield and mortgage rates to reflect stigma but did not specifically deal with the magnitude of uncertainty of remedial costs. Patchin also acknowledged that the major problem in environmental analysis is the lack of available market data. He suggested that more precise valuation techniques would be developed when market data become more available.

2.4 Stigma and the effects of 'time' and 'information'

Bell adopted the graphic used by Mundy (1992a) (Fig. 2.1) in which he illustrated the change in the value of contaminated real estate over time. The greatest loss in value is at the time when contamination is discovered, which can result in a property becoming unmarketable until the problem is investigated and the seriousness of the contamination is understood. The new knowledge can reduce the perceived risk, resulting in an increase in value. The next increase in property value occurs when the contamination is remediated. The graphic suggests that the loss in value will eventually decline to zero over time.

Although the graphic is useful, it does not take account of a potential increase in value as stigma decreases from, for instance, lower remediation costs or, at the end of the remedial period, when a regulatory authority confirms that no further action is required.

Mundy theorised how a loss in value can be quantified by discounting the present value of the lost income over the duration of the loss. In a subsequent article, he described further methods of quantifying the impact of contamination (Mundy, 1992b). He concluded that contamination could influence both the income generating capability of a property and the level of risk associated with it.

Weber (1998) discussed in detail the requirements of the Remedial Investigation/Feasibility Study for a contaminated site. He quantified the uncertainty in the remedial costs for a site in terms of probability density functions and introduced the use of Monte Carlo sampling for risk analysis. The risk analysis was based on the 'triplet theory of risk' that considers:

- What can go wrong?
- How likely is it to happen?
- If it does happen, what are the consequences?

These factors are illustrated in Fig. 2.2. The result is a probability distribution that provides a good graphical representation of the relative financial risk of a site. Similar types of probability distributions are often used by institutional investors for quantifying the risks of real estate development.

Weber recommended the use of a land development model, similar to the one described by Coughlin (1995), which uses Monte Carlo methods to estimate the present value of the property in its contaminated state, without the use of indemnifications or insurance. These tend to cloud the impaired value of the freehold (fee-simple) interest in the property. Although remedial costs were treated as a cost of development, Weber suggested that the uncertainty of remedial cost is very

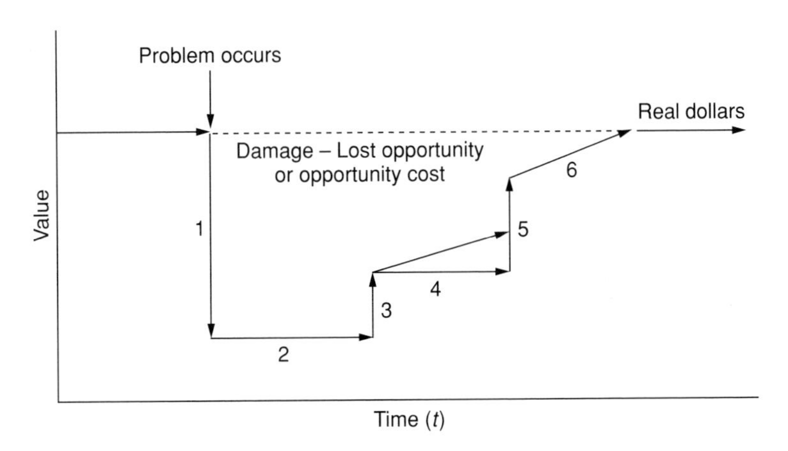

1 = Loss in value as a result of diminished marketability. Public becomes aware of the problem. How buyers and intermediaries perceive risk (unknown and dread factors).
2 = Duration. Time to understand relationship between hazard and risk.
3 = Amount of improvement in value resulting from knowledge (scientific) of hazard and effect knowledge has on perceived risk.
4 = Duration as hazard remains. Value may change as perceived risk changes.
5 = Increased value caused by removal of hazard (i.e. cost to cure).
6 = Stigma remaining after hazard removed. A period of uncertainty related to uncontrollable, involuntary, unknown, unobservable character of the hazard.

Damage—Related to opportunity. Lost opportunity is an opportunity cost measured by the diminished value over the duration of the event at a market rate.

Fig. 2.1 Changes in the value of contaminated real estate over time. Reprinted with permission from *The Appraisal Journal*, April 1992, Impact of Hazardous Materials on Property Value.

Analysis of Operable Units I–II

	Remedial Uncertainties						Joint Probability	Remedial Cost	Cumulative Remedial Cost	Expected Cost	Time to Closure (quarters)	Cum'ive Time (quarters)
I	**Tank and Lines**											
A	Only have to excavate & remove	40%	NFA (No Further Action)				40.00000% (col 1, row A)	$4 100	$4 100	$1 640	0.05	0.05
B	Samples required at tank ends	60%	99%	NFA			59.40000% (col 1×col 2)	$2 790	$6 890	$4 093	0.3	0.35
C	Contamination found-R A req'd	60%	1%	40%	NFA		0.24000% (1×2×3)	$3 000	$9 890	$24	1.25	1.6
D	Borings to determine extent	60%	1%	60%	90%	NFA	0.32400% (1×2×3×4)	$12 420	$22 310	$72	1.5	3.1
E	Additional investigation req'd	60%	1%	60%	10%	99.9%	0.03600% (1×2×3×4)	$5 050	$27 360	$10	2.5	5.6
F	Soil vapor extraction req'd	60%	1%	60%	10%	0.1%	0.00004%	$60 000	$87 360	$0	11	16.6
	Simulated Remedial Cost I:								$6 890			
	Simulated Time I I Cost A-E:											0.35
II	**Oil Drain Pit**											
A	Dig up concrete/dispose of soils	80%	NFA				80.00000% (col 1, row A)	$4 815	$4 815	$3 852	0.6	0.6
B	Regulators observe/require samples	20%	90%	NFA			18.00000% (col 1×col 2)	$2 275	$7 090	$1 276	1.25	1.85
C	Contamination found-determine extent	20%	10%	99%	NFA		1.98000% (1×2×3)	$8 360	$15 450	$306	2.5	4.35
D	Deeper borings req'd	20%	10%	1%	NFA 99.9%		0.01998% (1×2×3×4)	$5 220	$20 670	$4	3.125	7.475
E	PCE Found – Remediation req'd	20%	10%	1%	0%		0.00002% (1×2×3×4)	$1 000 000	$1 020 670	$0	18	25.475
	Simulated Remedial Cost II:								$4 815			
	Simulated Time II I Cost A-E:											0.60

Analysis of Operable Units III–IV

III Waste Oil Pit										
A Excavate tank and dispose	30%	NFA				30.00000% (col 1, row A)	$4 270	$4 270	$1 281	0.1	0.1
B Regulators observe/require samples	70%	40%	NFA			28.00000% (col 1×col 2)	$2 120	$6 390	$1 789	1.25	1.35
C Borings for additional tests	70%	60%	80%	NFA		33.60000% (1×2×3)	$5 870	$12 260	$4 119	2	3.35
D Deeper borings for additional tests	70%	60%	20%	99.0%	NFA	8.31600% (1×2×3×4)	$5 650	$17 910	$1 489	2.7	6.05
E Remediation of oil residue req'd	70%	60%	20%	1%		0.08400%	$50 000	$67 910	$57	10	16.05
Simulated Remedial Cost III:								$17 910			
Simulated Time III I Cost A-E:											6.05
IV Clarifier & Leach Field											
A Only have to remove clarifier	20%	NFA				20.00000% (col 1, row A)	$10 145	$10 145	$2 029	0.1	0.1
B Find contamination, further tests req'd	80%	80%	NFA			64.00000% (col 1×col 2)	$6 670	$16 815	$10 762	0.75	0.85
C Tests find dichlorobenzene	80%	20%	99%	NFA		15.84000% (1×2×3)	$8 650	$25 465	$4 034	2	2.85
D Tests find PCE	80%	20%	1%	100%	NFA	0.15984% (1×2×3×4)	$25 290	$50 755	$81	3	5.85
E PCE threatens drinking water	80%	20%	1%	0.10%		0.00016% (1×2×3×4)	$250 000	$300 755	$0	16	21.85
Simulated Remedial Cost IV:								$16 815			
Simulated Time IV I Cost A-E:											0.85

		I	II	III	IV	
Operable Unit:		I	II	III	IV	
Cumulative Cost		$6 890	$4 815	$17 910	$16 815	
Total Cost:		$43 640				
Time:		0.35	0.60	6.05	0.85	(quarters)
Maximum Time:		6.05	(quarters)			
Worst Case Scenario:		$1 476 695			25	
		Pessimistic	M. Likely	Optimistic	Forecast	
Value after Remediation (1000s)		$1 700 000	$2 000 000	$2 200 000	$1 950 000	
Simulated Present Value	$1 492 377					

Fig. 2.2 A method of assessing the risk from land contamination, in terms of the 'triplet theory of risk.' Reproduced with permission from the author.

likely to account for the greatest loss in value of a contaminated site and, as a result, can be the largest component of stigma.

The approach adopted by Weber in 1998 differed to that of Wilson (1994) in that an attempt was made to identify a continuous probability distribution that most closely estimated the remedial cost. The continuous probability distribution was sought so that the differences in remedial cost could be expressed probabilistically. It can also be used to extract the level of risk aversion in the local market since it makes use of a continuous probability density function.

Wilson also discussed what he referred to as indirect impairments, i.e. a detrimental effect on property value allegedly influenced by a nearby known environmental risk, such as a Superfund site.[2] In his view this was usually the most difficult analytical situation of all and one where popular perceptions are frequently not supported by the data. He also noted that part of the difficulty results from appraisers' tools not being useful for such an analysis. One tool that he suggests is useful is geostatistics that measure an assumed inverse relationship between distance and the effect of a variable.

Wilson's conclusions relative to uncontained impairments are that:

(1) Uncontained impairments present new legal and intellectual problems for the valuer, as the source of the problem (contamination) may not be on the subject property.
(2) There can be a very large spread in the possible costs of remediation, leading to greater risks and uncertainty. Often the risks and uncertainties result in value offsets much greater than the estimated cost of dealing with the facility.
(3) Standard appraisal techniques will not work adequately with most uncontained impairment situations as an engineering impaired model would need to be employed and such expertise may be beyond an appraiser's capability.

Weber (1996, 1998) agreed with Wilson in acknowledging the need for appraisers to become more knowledgeable about the technical and scientific factors of environmental site assessment and remediation in order to better critique environmental site assessments that they are so dependent upon.

Simons (1998) noted that there are a number of routes by which a brownfield site can be identified:

[2] Superfund is a revolving trust fund established by the US Congress in the Comprehensive Environmental Response, Compensation, and Liability Act of 1980 to cover the cost of cleanups.

- A private developer or not-for-profit developer inadvertently stumbles across one while assembling a property for development.
- A seller or potential buyer seeking financing discovers that contamination exists when the property is marketed for sale.
- A developer actively seeks such properties, hoping to buy them at a reduced price, remediate real or perceived contamination, and later sell the properties at or close to market value for 'clean' property.
- Government agencies or financial institutions become the owners of last resort or have an opportunity to acquire a brownfield or finance one.

Simons, who has a planning background, went on to observe that experts cannot agree on what a brownfield is and noted the lack of expertise on the part of valuers and their ability to undertake brownfield valuations. He concluded that instead of providing valuations that reflect the impact of contamination, nearly all appraisers provide a value 'as if clean'. He concluded that without the valuation reflecting the true state of the land, which would provide value net of remediation cost and stigma, banks cannot ascertain a property's value for consideration in the loan-to-value ratio, and loans become even more difficult to obtain.

Jackson (2000) expressed the view that the question of whether or not contamination continues to affect the value and price of properties after remediation is important for valuers, as well as for those involved in contaminated property transactions, and also for courts that have the task of considering damages due to contamination. The valuation of contaminated and previously contaminated properties will have a different focus, depending on the date of value, if adverse impacts due to contamination are known to dissipate subsequent to remediation. Therefore, the 'stigma' effect attributable to land contamination may be seen in a number of different ways:

- *Initial stigma*: the price adjustment over and above any estimated cost to correct what may be assigned to a property immediately if the presence of the contaminants is known but before any comprehensive site investigation or risk assessment has taken place. This is closely associated with Mundy's and Wilson's views on 2X or double counting through both allowing for costs and in upwards revisions of investment yields.
- *Pre-remediation stigma*: the price adjustment that may be expected in respect of a site that has been fully investigated and for which a risk assessment has been prepared – a situation that may be seen

as a 'sale with information' under Part IIA of the Environmental Protection Act 1990.

- *Post-development stigma*: any price adjustment that remains after the site has been redeveloped, i.e. a discount against prices achieved on similar developments in equivalent markets. In practice, with a well-thought-out development in which the site remediation has been properly planned and executed and where the information has been properly communicated to the purchaser(s) and/or investor(s), this should be negligible or incapable of being measured.
- *Future stigma*: any price adjustment on second or subsequent developments following redevelopment, which may be seen as the fear of something going wrong. In practice this should not exist in respect of a good redevelopment, especially where the information regarding the site treatment has been properly recorded and retained for future transactions; indeed there is the possibility that buildings constructed on a properly remediated and documented site may attract a 'premium value' over buildings on similar developments where less is known about the site history.
- *Proximity stigma*: any price adjustment attributable to the location of a potentially contaminative activity close to the subject property, or to the presence of known contamination on adjoining land. This is the most difficult type of stigma to deal with as it is outside the control of the land owner, who may have to rely on a regulator to take action against the polluter, or embark upon a costly civil action in tort.

Jackson (2000) noted that some courts view property value diminution due to contamination as a temporary condition, under the assumption that remediation will 'cure' the diminution. On this basis compensation would be based on the cost of remediating the property or the diminution in market value, but not both. Jackson's research looked at different valuation methods in an attempt to evaluate whether previous contamination did have an effect on sales prices. Using two procedures, sales comparison analysis and multiple regression analysis with two model specifications, he concluded that the previously contaminated properties did not sell at prices that differed from their uncontaminated comparables.

In looking for a method of estimating the economic damage caused to property values due to loss of marketability, rentability and stigma, Mitchell (2000) looked firstly at direct damages and secondly at indirect damages (stigma). He described indirect damages as those relating to the increased risk of further economic loss associated with the contamination or other impairment or defect, and the effects that this may have upon marketability, rentability, mortgageability, insurability, income, and value of the property *after* remediation and confirmation of such issued by an appropriate authority.

Mitchell gave an example where serious contamination was found rending the property unsaleable. He assumed that the cleanup would take four years but that the costs will be met by insurance cover. He presented a range of scenarios where there was contamination with and without stigma; with immediate cleanup; one year delay; minor construction defect; and a worst case scenario. He concluded that most of the economic damage would occur during the cleanup period with only minor loss due to subsequent stigma.

2.5 Summary

There are many similarities in the valuation approaches that have been adopted by practitioners in the United States and the United Kingdom. In some instances the methods employed have been highly theoretical, for example based on almost arbitrary adjustments to yield rates, whilst others have been based on more factual information, such as the cost to correct the damage caused to the land. Even then questions may remain as to what should be taken into account as a 'real estate' cost when preparing the valuation, or excluded as a 'business' cost.

A number of different techniques have been discussed in this chapter, some of which are more suited to the valuation of investment properties, whilst others are more appropriate for use in owner occupation valuations or for development appraisals. Valuers embarking upon the valuation of contaminated properties will have to decide which method is most suitable for the type of valuation proposed and may also be well advised to compare the results obtainable from two or three different techniques. That is, of course, always assuming that they have enough information available to them to undertake the valuation in the first place and they feel competent in undertaking the valuation of a property affected by the presence of contamination in the land.

In both countries a significant problem is the lack of transparency attaching to property sales. Even in the United States, where transaction data are generally more accessible than in the United Kingdom, the parties to a transaction may not wish to place on public record the allowances made in respect of contamination and thus will limit the publicly available information with a confidentiality agreement. The presence of contamination can affect not only the 'value' of land but can actually go so far as to render it unsaleable in normal markets, or unmortgagable through conventional lending sources. This might apply even if the property were in full commercial use at the time of the valuation, especially if the site happened to be the subject of regulatory action.

Valuers need to take care when undertaking the valuation of land affected by contamination but, generally speaking, the more information they have available to them, the greater the accuracy of the valuation. It should also go without saying that the valuer's report should describe the information on which the valuation has been based and set out any misgivings that the valuer might have on the accuracy or validity of that information.

Checklist

- Ensure that the purpose for which the valuation is required is clear and set out in writing, e.g. does it clearly state whether or not the site is to be redeveloped immediately or in the foreseeable future?
- If the site is to be redeveloped, the potential remediation costs should be accounted for in the valuation, unless the instructions and report specify that these should not be taken into account.
- Potential risks to adjoining properties and possible legal liabilities to third parties should be identified and reported upon.
- Extending an existing building may involve a need to remediate the site, even for continued owner occupation, for example to comply with health and safety regulations for construction workers.
- The valuation report must clearly state the assumptions, including remediation costs, which have been used when preparing the valuation.
- The valuer should consider whether any 'stigma' exists before redevelopment and whether the property is likely to be stigmatised after completion of the development, including 'proximity' stigma from adjoining properties/activities.
- If using any comparisons when preparing valuations of land affected by contamination, be very careful that they are in fact 'comparable' and state the information source.
- Consider whether or not the presence of contamination will render the property unsaleable or unlettable.

Chapter 3
Barriers to Redevelopment

3.1 Introduction

In spite of the fact that Government policies are firmly in favour of redeveloping PDL in preference to 'greenfield' development, there are still a number of developers, albeit probably reducing in number, who are reluctant to develop on previously developed land. The reasons they give are many and varied but generally fall into one of the following categories:

- fear of the unknown
- regulatory controls
- delays
- increased costs
- stigma

The reluctant developer may also cite a combination of two or more of these as the reason for considering redevelopment as being 'too risky'. These concerns are perhaps greatest amongst the developers of residential properties, where a higher risk of harm to human health might be perceived. This chapter considers the concerns arising under each of these five categories and suggests ways in which they might be overcome, or at least reduced.

3.2 Fear of the unknown

This is an important barrier to redevelopment and one that is often difficult to assess in financial terms. Therefore it is essential that the intending developer takes full account of current and previous land-use activities, which often results in the question being asked 'but how far back should I go?' Strictly speaking the answer should be 'to whenever the site was last a greenfield' but, in practice, information relating to the earliest land uses in some of our industrial towns and cities may be shrouded in the mists of time. So all the diligent investigator can hope to do is to go back as far as reasonably possible, using information from a variety of data sources.

The site investigation is all-important and, regardless of the form and type of the proposed development, it must commence with a thorough understanding of the historical context of the site and its surroundings. Syms (2002, Chapter 5) describes the process of site investigation from the perspective of what the 'non-specialist' reader should be looking for in a site investigation report and makes the point that:

> 'Any intending developer, or planning officer, should be extremely wary of a site investigation report that does not contain an historical study and, unless satisfied by further enquiries, probably would be well advised to reject the report.'

Given that the CLEA model has been published since this was written (see Chapter 1), it is perhaps appropriate to add a further caution:

> If the intrusive investigation has been undertaken since December 2002 and if the ICRCL 59/83 trigger concentrations have been used as the basis for the risk assessment, then the report should probably be rejected.

Petts *et al.* (1997) provide a very useful table which details the 'Key stages in risk assessment and the consequences of poor investigation'. The authors provide several very pertinent examples of the consequences of poor site investigation, including:

- Inadequate site history leads to failure to identify potential contaminants, putting the health and safety of the investigation team at risk and may result in prosecution.
- Failure to recognise that the regional direction of groundwater flow may have been modified by local features such as an extraction well, waste tip or deep foundations, may result in the inappropriate siting of sampling wells.
- Presence of protected species not identified, resulting in damage during investigation and subsequent prosecution.

- Trial pit or borehole penetrates low-permeability layer, allowing contaminants to reach underlying aquifer, landowner prosecuted and sues investigator.
- Soil samples left in the hot sun, allowing degradation, or placed in inappropriate containers, leading to loss of components and/or reaction with the container.
- Quantity of data insufficient for valid comparison according to guidance attached to guideline values.
- Laboratory employs inappropriate sample preparation method, with the result that volatile or degradable components are lost.
- Inappropriate risk estimation methodology, or inappropriate assumptions used, resulting in underestimation of the risks.

All of these will lead, at best, to uncertainties regarding the quality of the site investigation and may well result in prosecutions, or in members of the site investigation team being sued for professional negligence. If undetected, inadequacies in the site investigation may have even more wide-reaching consequences, such as the developer paying too much for the site, to the extent that previously undiscovered contamination renders the development project commercially unviable. Even worse is the potential situation where, as a result of the poor site investigation, the developer inadvertently becomes a polluter under the Part IIA legislation, for example by releasing hitherto contained contaminants as the result of piling or other engineering operations, thereby causing harm to the site workforce or to the wider environment.

The risks described above may be seen as being more than enough to discourage the reluctant developer of PDL, so how can they be mitigated? Firstly, the developer needs to know the types of contaminants that might be present in the ground and in what sorts of quantities and/or concentrations. One very useful tool for obtaining this information has been the *Industry Profiles* published by the former Department of the Environment. These provide information on the processes, materials and wastes associated with individual industries, as well as information on the contamination that might be associated with specific industries. Factors that affect the likely presence of contamination and the effect of mobility of contaminants are also described and guidance on potential contaminants is provided in the profiles. The profiles are not definitive studies, nor can they be expected to cover all types of sites used for different industries especially as working practices tend to change over time, but they introduce some of the technical considerations that need to be borne in mind at the start of an investigation for possible contamination.

The DoE *Industry Profiles* comprise a total of 48 volumes (see the end of the book for the complete list) and were prepared from research undertaken by a number of different consultants commissioned by the Department of the Environment. Several of the profiles are now out of print and one purpose of this book is to summarise much of the information from the profiles in a single volume. Nevertheless, it must be stressed that the industry-based information contained in Part B is only a précis of the information obtainable from the profiles and other sources, although it should be sufficient for the purpose of identifying those contaminants that are likely to be present on sites used for different industrial purposes.

Even when the investigation has been carried out to recognised standards of industry good practice there will remain uncertainties, simply because that is the inherent nature of ground investigation. These uncertainties may result in cost overruns in remediation or in residual exposure to the potential for contamination left in place, either deliberately or otherwise. The financial risks of these types of uncertainty can be managed using one or more of a range of environmental insurance solutions (see later).

Second, after investigating the site and defining any hazards that might exist, two basic questions remain (LaGrega *et al.*, 1994):

(1) To what extent should the site be remediated to protect human health and the environment (i.e. 'how clean is clean')?
(2) How should this level of protection be achieved (i.e. what is the 'best' remedy)?

LaGrega *et al.* referred to these two questions collectively as 'selection of remedy' and express the view that 'rarely is the decision self-evident', as several important factors compel the use of an open-ended decision-making process that promotes full participation by interested parties. The objective of such an approach being to progress to a conclusion whereby the fundamental purpose is realised, namely that the site is remediated in the manner most appropriate to the proposed use and associated risk factors.

This author has long held the view that engineers are not necessarily the professionals most suited, in isolation, to advise on the selection of remediation methods. A team approach is needed, which will include the letting and/or selling agent, the investment surveyor or mortgage advisor as appropriate and the lawyers, both from the environmental perspective and in respect of sale and transfer. It is also important to involve the local community, especially if the remediation work is likely to involve noise, dust or odour, or entails large numbers of heavy vehicle movements to and from the site.

The concerns of these individuals and groups will, inevitably, differ, as too will their values. The developer will be seeking to achieve a remediation solution that is cost effective and impacts least on the profitability of the project but this will be of little concern to the local community group, who will be more concerned to ensure that a safe solution is found with the least disturbance during the period in which the works are undertaken. The engineer may propose a solution that meets the developer's criteria but for some reason or other, for example the need for mechanical venting systems or post-completion monitoring, may not be acceptable to the letting agent on grounds of lettability or to the investment surveyor because of potential impact on the investment yield.

From the developer's perspective the discussion regarding contaminants found on previously developed land often comes down to a 'make or break' decision as to whether or not to continue with the project or to abandon it (see Fig. 3.1). With some forms of immobile contaminants it may be possible to accept that they will remain in the ground after

Fig. 3.1 The site of this mixed industrial and car sales development had previously been a rail goods yard, including cattle transfer station, and town gasworks. It was also extensively fly-tipped. The developer commissioned a very detailed site investigation before finalising the design and remediation works.

the new development has been completed, either distributed across the site or contained under some impermeable cover, requiring consideration to be given only to providing protection to building workers against short-term exposure risks. The acceptability, or otherwise, of this type of approach will then depend on the attitude of the investing institution or upon the mortgage provider.

In all other situations, where some form of remedial action is required, the developer will want to know whether this can be achieved on site, either in-situ or ex-situ, or if off-site removal of the contaminated soil will be required, for either treatment or disposal. Many developers will prefer the off-site disposal option to any on-site treatment or containment, as it is seen as a 'once and for all' solution to the problem, removing any uncertainty that may attach to on-site methods. However, there are environmental risks attaching to the off-site disposal option, both in terms of the risks associated in transporting large volumes of contaminated soil and in ensuring its proper disposal.

Finally, in terms of 'fear of the unknown' is the issue of possible contamination to ground or surface waters. During the course of research undertaken for the Joseph Rowntree Foundation (Syms, 2001), it became very clear from talking to developers that the problem they feared most, in redeveloping PDL, was having to deal with ground or surface water contamination. Soil contamination they could cope with but contaminated water, with the implications that contaminants could migrate over considerable distances, was another matter. This fear of water contamination is perhaps the greatest barrier of all to the redevelopment of some sites. Again this is the type of problem that lends itself to environmental insurance, subject of course to information and an ability to analyse and understand the potential risks and exposure over time.

3.3 Regulatory controls

Some concern exists among developers as to which authorities are responsible for regulating redevelopment. At least one well known housebuilder has made the mistake of agreeing a remediation plan with the Environment Agency, only to find that the proposed works were not acceptable to the local authority's environmental health officer, who recommended refusal of the application, resulting in a planning appeal. As stated in Chapter 1, current Government policy is that land contamination should be dealt with as part of the planning/development process. This applies except where 'significant harm' is being caused, or there is potential for such harm to be caused, or the pollution of controlled waters is taking place or is likely to occur.

It follows therefore that regulators likely to be most concerned with any remediation proposals will be the local authority's planning and environmental health departments, although the Environment Agency will be a statutory consultee. Current policies and guidance to local planning authorities in respect of planning applications for PDL where there is the potential for contamination to be present are set out in the following pages.

3.3.1 *Land contamination planning policies and guidance to planning authorities*

As of summer 2004, the guidance issued by central government to local planning authorities, as to the redevelopment of land that might be affected by contamination, is contained in Planning Policy Guidance Note 23, *Planning and Pollution Control* (PPG23) (DoE, 1994). However, this document has been the subject of revision for some time and the section on Waste Disposal has been revised and re-issued as PPG10 – *Planning and Waste Management*.

In February 2002 the Office of the Deputy Prime Minister (then DTLR) initiated a consultation process on a new Planning Technical Advice Note, *Development on Land Affected by Contamination*, which was issued as a consultation draft. In July of the same year the ODPM also commenced consultation on guidance to replace the remaining parts of PPG23. This guidance was intended to update those parts of PPG23 dealing with pollution control, air pollution, water pollution and climate change.

The consultation draft Planning Technical Advice Note *Development on Land Affected by Contamination* took into account the statutory arrangements for contaminated land introduced on 1 April 2000 (EPA 1990, Part IIA) and applied only to England. It set out the responsibilities of the Parties in the Development Process, including:

'Role of the developer

Where development is proposed, the primary responsibility for safeguarding land and other property, including neighbouring land, against any risk from contamination remains with the owner. It is the responsibility of the developer to ensure that a development is safe and "suitable for use" for the purpose for which it is intended. The responsibility for determining whether land is suitable for a particular development rests with the developer. In particular, the developer is responsible for:

- Determining whether any proposed development will be affected by contamination and whether it will increase the potential for contamination on that site or elsewhere.
- Satisfying the local planning authority that any contamination can be successfully remediated with the minimum adverse environmental effect to ensure the safe development and secure occupancy of any site.

Role of the local authority

The local authority has the following roles:

- In most cases, the environmental health departments of local authorities are the enforcing authorities for the new contaminated land regime.
- Local planning authorities are responsible for the control of development, and the enforcement of conditions to ensure that development is appropriate in relation to all the relevant circumstances of the site and any contamination, that land is remediated in the course of development to an appropriate standard for its intended use, and that it is properly maintained thereafter.
- Building control – a developer or builder intending to construct a new building is required by law to obtain building control approval which includes the requirements to protect buildings from the effects of contamination. There are two types of building control providers – the Local Authority and Approved Inspectors (a private sector alternative).

Role of the Environment Agency

The Environment Agency has certain responsibilities under the new contaminated land regime, specifically in relation to "Special Sites" but, more generally, the Agency carries out technical research and publishes scientific and technical advice relevant to land contamination. The Agency is a consultee on certain planning applications for development on land which may be affected by contamination (GDO 1995 – proposal which could lead to increased industrial discharge into a river or estuary and proposals within 250m of notified landfills).' (DTLR, 2002, paras 17–19)

The consultation draft noted that any land may be affected by contamination, including 'greenfield' as well as previously developed land, and that contamination can arise from natural sources as well as from man-made sources. It also stressed that policies in development plans should outline the consideration that will be given to land that may be affected by contamination, recognising the uncertainties that are inherent in identifying, investigating and remediating such land.

The consultation draft went on to stress that 'where practicable, informal pre-application discussions between the LPA and prospective applicant should take place', as they can be helpful in identifying whether the land may be affected by contamination and the implications of any contamination. It also emphasised that LPAs should pay particular attention to the condition of land where the proposed use of the land would be more vulnerable to any past contamination; or where the current circumstances or past use of the land suggest that contamination may be present, or where it has other relevant information. Full account should be taken of whether the proposed use or development for which planning permission is being sought is capable of being adversely affected by contamination, if any (DTLR, 2002).

In order to address these issues, LPAs will need to examine their own historical records, in order to ascertain whether they contain any information about possible contamination that might affect the development and they will need to be provided with information by the prospective developer. As a minimum the information required from developers of PDL where contamination might be an issue should comprise a desk study. In many situations, this may be sufficient to satisfy any concerns that the planning officer may have, as not all urban land uses will have included industrial activities with the potential to cause contamination, and thus allow a planning application to be determined.

When the desk study establishes that contamination is likely then the study by itself may not provide sufficient information to support a decision to grant planning permission. A full site investigation and risk assessment may then be required where the evidence available to the authority justifies this. Whatever the circumstances, the information required should be that reasonably necessary to enable the LPA to decide whether to grant or refuse planning permission. Desk studies and other information could form part of any environmental impact assessment which may be required (DTLR, 2002).

Subject to the LPA being satisfied with the information provided by the developer and from its own investigations, planning permissions should include appropriate conditions or obligations for development on contaminated sites. These should cover the possibility of discovering previously unsuspected contamination during the development process, site investigation, remediation. In addition, they should deal with the provision of effective maintenance and monitoring supported by sufficient reporting and record keeping at all stages, so that the permitted development is or can be made compatible with the circumstances of the site (DTLR, 2002).

The changes to the rest of PPG23 were drafted on the basis of arrangements in force on 1 April 2002 and took account of:

- The setting up of the Environment Agency under the Environment Act 1995.
- The new pollution control framework under the Pollution Prevention and Control Act 1999 and the Pollution Prevention and Control Regulations 2000.
- The national Air Quality Strategy and the system of local air quality management under Part IV of the Environment Act 1995.
- The Groundwater Regulations 1998, which prohibit the disposal of listed substances to land without prior authorisation from the Environment Agency, and enhance the controls over direct discharges to groundwater in the Water Resources Act 1991.

As with the draft guidance in respect of land affected by contamination, this draft guidance applied only to England and it advised that (ODPM, 2002a):

- Any air or water quality consideration that relates to land use and its development is capable of being a material planning consideration.
- The planning system has an important role to play in determining the location of development which may give rise to pollution, and in ensuring that other developments are, as far as possible, not affected by major existing, or potential sources of pollution.
- The controls under the planning and pollution control regimes should complement rather than duplicate each other.
- Developers will find it helpful to hold early discussions with both the relevant local planning authority and the relevant pollution control authority and other relevant bodies on the proposed development.
- Where possible *parallel* submission of applications under planning and pollution control procedures should be encouraged.

The consultation process in respect of the changes to PPG23 ended in October 2002 but implementation of the changes to planning guidance have been delayed due to the wide-ranging review of town planning procedures that followed publication of the Planning Green Paper (DTLR, 2001) and the Government's response to the ensuing public consultation (ODPM, 2002b), see Chapter 6. It is expected that these two aspects of change will be brought together in a single short, sharply focused, Planning Policy Statement, with two annexes covering the issues dealt with in the two consultation exercises.

3.3.2 *Other 'regulatory' concerns*

One concern expressed by developers and their advisers, interviewed as part of the 2001 Joseph Rowntree Foundation study into the problems surrounding the return of 'brownfield' land to beneficial use (Syms, 2001), was the lack of any national standards for the remediation of contaminated soils. 27% of interviewees considered 'set standards' to be 'essential' and 28% regarded them as being 'very important' (Syms, 2001, p. 47). The argument in favour of set remediation standards is that it provides a 'level playing field' whereby developers, who may be bidding against each other for a site, know what has to be achieved and can budget accordingly. The argument against is that every site is individual, with contaminants contained in differing media and in a variety of environmental settings so, instead of prescribing standards, the UK approach has been to produce the CLEA Guideline Values, enabling developers and their consultants to assess what is required on a site-specific basis according to the type of development proposed.

Developers are also concerned that, having acquired a site, it is then designated by the local authority as 'Contaminated Land', or by the Environment Agency as a 'Special Site' under Part IIA of the Environmental Protection Act 1990 (see Chapter 1). Whilst it is certainly the case that a developer could become a Class A person under the legislation by 'knowingly permitting' the continuation of a contaminating activity, or by involuntarily creating a significant contamination pathway as the result of engineering or construction operations on the site, that is certainly not the intention of the Part IIA legislation, under which voluntary remedial action is encouraged in preference to compulsion. Therefore, the authorities are more likely to want to work with the developer in finding a solution to the problem than in going along the road of regulatory action.

As at February 2004 only 71 sites had been determined as 'Contaminated Land' of which 17 were 'Special Sites'. Of these 71 sites, identified in the first three years and 11 months of the legislation being in force, only 3 Remediation Notices have been issued and 23 Remediation Statements registered. It should be stressed that not all of the sites will require physical remediation, in the form of removing or containing contaminants or changing the nature of the soil, as the term 'remediation' under the legislation also includes site investigation and monitoring.

The Environment Agency has not analysed the determinations so as to assess the total area of land affected by the determinations but, assuming that the sites determined to be 'Contaminated' or 'Special' have an average area of five hectares, and some are significantly smaller than this, the area involved might amount to a total area of 355 hectares. In 1997 this author stated 'it would appear that the extent of land in Britain affected by contamination which could cause "significant harm" may be around 2800 hectares (6900 acres)' (Syms, 1997b, p. 289). Admittedly there is still more land in the pipeline but, at the present rate of determination of 'Contaminated Land' or 'Special Sites', it will be several more years before this estimate (which was regarded by some as unrealistically low) will be reached, if it is ever achieved.

Although the area of land (and number of sites) identified so far as having to be dealt with under the legislation is small, the legislation has had the effect of encouraging some land owners to remediate or redevelop land without waiting for action to be taken by the regulatory authorities. This was one of the objectives of those responsible for drafting the legislation and it serves to demonstrate that genuine developers probably have little to fear from Part IIA.

3.4 Delays and increased costs

Environmental liabilities and risks have jeopardised many property and land transactions over the last few years, and may justifiably be perceived as barriers to redevelopment. However, with careful thought, many of these can be provided for in the development appraisal. Increasingly, buyers are concerned about taking on historic liabilities associated with pre-existing contamination or suspicions about contamination, and 'selling with information' is one way in which polluters can potentially be excluded from liability under Part IIA of the Environmental Protection Act 1990. Equally, sellers are anxious about retaining liabilities after disposal of a site, as it may have unquantifiable financial implications for years to come, which may need to be noted on a company's report and accounts. Lenders are sensitive to the credit risk exposure should the value of the investment fall, if the land or property is compromised by environmental liabilities, as the collateral for the loan may be greatly diminished. These concerns are relevant irrespective of whether any remedial action has been taken. The intending developer therefore needs to make a careful assessment about the quality of the information available at the time of site acquisition, environmental reports and the like, and the extent to which liabilities are transferred or retained.

Recent developments in the UK in policy implementation, regulations and the publication of a significant body of professional guidance have done much to create a positive market for brownfield and contaminated land. Nevertheless, both short-term and long-term fears and uncertainties continue to exercise the minds of all parties involved in such transactions. In some situations the developer's preferred remediation solution may not be feasible, possibly due to environmental constraints (such as noise, dust, vibration or vehicle movements) affecting neighbouring properties, or the inability to obtain either a Waste Management Licence or a Mobile Plant Licence for the processes involved. Changing the remediation methods may increase the development cost and also cause delay.

Recently, and in parallel with a greater interest and need, through market pressure, to transact and redevelop brownfield sites, the insurance market has reacted to provide an increasing range of facilitative solutions.

3.4.1 Insurance solutions

Based on an understanding of the technical risks gained from environmental reports, Environmental Insurance Solutions can be provided, typically, for up to 10 years on a 'claims made' basis. This means that a claim has to be made within the stated policy period. They can be transferable/assignable on change of ownership of the land and can even be written so as to allow for later subdivision of the land in question and to protect the original landowner from any residual liability. Various structures and 'products' can be sourced from a number of UK-based insurers to respond to a wide range of associated financial risks including:

- Contaminated Land Insurance for historic contamination risks and exposures.
- Operational Risk for on-going processes and operations on the site.
- Legal Indemnity to provide insurance backing for environmental indemnities and warranties.
- Remediation Cost Overrun (Stop-Loss) to control clean-up costs.

- Secured Loan to provide lenders with an alternative to going into possession of environmentally suspect or contaminated land.
- Contingent Reserve as an insurance alternative for businesses who would otherwise hold reserves against potential environmental liabilities.
- Contamination and Land Development Solutions to protect developers against erosion of investment returns by contamination.
- Portfolio Options for more than one site.
- Risk Financing for protecting against future liabilities and remedial action costs, where insufficient information is available to provide immediate and full risk transfer.

The insurance is usually for contamination, as specifically defined in the policy wording, that originates on-site and stays on-site; originates on-site and migrates off-site; or which originates off-site and migrates on-site. Cover can be provided for contamination that renders the site unsuitable for its use whether or not legal action is taken, for example by a neighbour or by an environmental regulator such as the Environment Agency. Where clean-up costs are incurred under the direction of the regulator these can be met by 'Clean-up Liability' cover. There is full civil liability protection against actions by third parties and legal defence costs are covered for cases of criminal liability (it is not possible to insure against criminal liability as such). Personal Liability of directors and officers arising under common law or statute can be insured.

Environmental insurance has developed at pace over the last 10 years and in the US has become very much a standard part of many developers' and funders' approaches to managing brownfield risks. Similarly the UK has seen the market virtually double each year over the past 5 years. Increasingly environmental insurance is being used to protect owned property and property portfolios as well as providing a valuable risk management tool in land and property transactions.

3.4.2 *Further site investigations*

The need to undertake further, or supplementary, site investigations may come about for a number of reasons, even when the original investigation has been well designed. For example, the risk assessor acting on behalf of the insurance underwriter may require more information regarding certain types of contaminants or parts of the site. Laboratory work may reveal elevated concentrations of some contaminants that were not envisaged following the historical study and the site walkover. This may lead to the need for a more intensive investigation of part of

Fig. 3.2 Even with a well planned investigation, it may sometimes be necessary to return to a site for supplementary investigation to be undertaken to confirm findings or to resolve anomalous situations.

the site, so as to determine whether the contaminants originated from within the site or had migrated into the soil from adjoining land (see Fig. 3.2).

Further site investigations may also be required in situations where the site was not fully accessible at the time of the original investigation. This would include situations where one or more buildings on the site were still occupied and where a vendor was reluctant to allow intrusive investigations through floors and yard areas until such time as contracts for sale have been exchanged. Prudent developers would be well advised to ensure that the contract is so conditioned as to allow re-negotiation of the price and/or the completion date if anything untoward is discovered during the supplementary investigation.

3.4.3 *Other delays that may impact on timescale*

Developers have expressed concerns about the potential for delays that might be caused by requirements to undertake archaeological digs, or at least to make the site available for such digs to be undertaken by others (Syms, 2001, pp. 43–4). If the need for an archaeological dig is known well in advance, for example, if the site crosses or is adjacent to an ancient city wall, the delay and indeed the cost of the dig can be factored into the development appraisal and land price. However, in situations where there is no condition requiring an archaeological investigation to be carried out but where human remains or artefacts are discovered during site preparation works, this is not so easy. The best the intending developer can hope to do is undertake a comprehensive historical study of the site and assess the potential for such finds to be made, then make appropriate provision in the development appraisal, including if possible the ability to renegotiate on land price. Consulting the county archaeologist or the archaeology department of a local university will help to reduce or even remove this barrier to redevelopment.

Another complaint often made by developers concerns the unwillingness of some local authorities to consult before a planning application is submitted. In the planning Green Paper (DTLR, 2001), it was suggested that planning discussions between applicants and local authorities would 'be encouraged, with local planning authorities being able to charge fees for such discussions so that they do not constitute a drain on resources. Such fees [would], however, have to be set at such a level as to ensure that they do not discourage applicants from early discussions' (Syms, 2002, p. 311). This idea has not been taken forward into the draft Planning Bill, possibly due to fears that charging fees for advance consultation may be perceived as selling planning permissions. The Office of the Deputy Prime Minister is consulting further on whether or not fees should be charged for advance consultations (see ODPM, 2002b, para 85).

3.5 Stigma

The question of stigma has been considered in the valuation context in Chapter 2 and, in particular, the different types of stigma that might exist according to whether a property is simply being valued, is about to be redeveloped or has already been redeveloped. The intending developer needs to be conscious of the potential impact of stigma before acquiring the site and should especially consider whether there will be any residual stigma after the development has been completed, as this will affect both the ability to fund the development and the development appraisal itself. This section looks at how developers may best assess the possibility of stigma and then reduce its impact.

3.5.1 *Position in the market*

The intending developer will need to consider the position of the proposed development relative to the local market. This means taking account of whether the development will complement what already exists, to the extent that it has the effect of lifting the market generally, or whether it is so much at odds with existing properties in the vicinity that it fails to attract buyers or tenants, or at least at commercially viable figures. That is not to say that developers should not contemplate undertaking developments that are significantly better than existing properties; on the contrary, without such entrepreneurial activities urban regeneration would not take place. Nevertheless, developers do need to pay particular attention to dust, vibration and odours from any industrial activities that may remain in the area.

Locating a new residential development, new build or conversion, next to an engineering works with heavy presses operating 16 hours a day would probably not be very successful. The units may sell but possibly at a significant price reduction when compared to similar developments in locations not affected by the proximity of an engineering works or similar neighbour. Larger developments may be able to create their own environments but may still be stigmatised to some extent by their surroundings, for example if the road access is through a run-down area. Good design will help to reduce stigma effects in many instances, especially if it provides the owner or occupier of the property with a feeling of being in a safe, comfortable environment.

3.5.2 *Communicating information*

Perhaps the most important aspect of removing the stigma barrier to redevelopment is that of communicating information regarding the site history, remediation works and any residual risks, to prospective purchasers, tenants and investors. Some developers might, quite understandably, be nervous about disclosing such information but, in the view of this author, openness is the best course of action. It is far less likely to lead to problems for the developer than say, the purchaser of a new house hearing from a local, shortly after moving in, 'Oh you live there do you, on the old cyanide works' – especially if cyanide had never been used on the site.

Having said that the best policy is to disclose information relating to the past history of a development site, the way in which that information is communicated is of fundamental importance. Scientific terminology such as 'the cancer risk has been estimated as 10^{-6}' is unlikely to instil confidence. Similarly, it may not be appropriate to display photographs of the site prior to redevelopment in prominent positions on the walls of the sales office but it may be appropriate to have these available and for the sales particulars to make a brief mention of the site history, concluding with the words 'a brief summary of the site history and the site preparation works are available from the sales team; fuller reports may be made available on request'.

Briefing the site staff on how to handle enquiries about site histories is important. It does little to instil confidence if the sales and construction personnel are only able to respond that they do not have any information or knowledge, especially when the development is obviously situated in a former industrial area. The best example the author has personally experienced, in terms of communication from the site team, was on the Wilcon Homes Eastern 'Hallywell Place' development at East Beckton in London, where the site manager was fully briefed in respect of the site history and the remediation, to the extent of having samples of the geo-synthetic clay liner available for inspection in the site office (see Syms & Knight, 2000, pp. 69–73).

3.6 Overcoming the barriers

Many of the barriers that might initially deter a developer from embarking upon a development on previously developed land can be successfully overcome. It requires a thorough understanding of the problems that might be encountered and careful planning as to how they can be overcome. Fear of the unknown is probably the most difficult barrier to overcome and, potentially, the one that might lead to the greatest financial errors in either under or over-estimating the cost of overcoming problems in the ground.

As mentioned above, specialist environmental insurance can be used to protect against cost overruns. However, insurance is not a substitute for a lack of investigation or design of the remedial works. It can be a particularly useful and facilitating tool where the remediation is being funded in full or part by some form of finite fixed grant.

At the end of the day, however, it all comes down to two things – the developer's attitude towards risk and appointing a professional team with the necessary skills to ensure a successful development.

Checklist

- A desk study, providing information as to past uses of the site, is essential prior to commencing a site investigation.
- The presence of an aquifer under the site must be clearly identified prior to undertaking trial pit or borehole investigations.
- Refer to Part B, or the appropriate industry profiles, for further information on sources of contamination and design the laboratory testing suite accordingly.
- Assemble an experienced professional team to advise on any remediation work that may be required and upon the selection of appropriate methods or treatments.
- Removing contaminated material for off-site disposal is not necessarily the most environmentally attractive method and may not be the most cost-effective solution. It is not risk-free.
- Consult with the planning officer and other regulators at an early stage.
- Do not assume that a greenfield site is uncontaminated.

Chapter 4
Recording Land Condition

Judith Lowe

4.1 Introduction

The previous chapters have provided the context for the reuse of previously developed land. They have addressed key aspects of appropriate valuation of such land, factors inhibiting development and the legislation and policies affecting PDL. Later chapters will consider complex legal issues and town planning issues, all of which may impact the reuse of land. To assess the implications of these for any individual site, it is essential to have information on the condition of the land, particularly in relation to contamination and other physical legacies of previous uses.

This chapter discusses information on land condition and current initiatives to improve the consistency and quality of the presentation and use of information about individual sites.

4.2 Information on land condition

Information on the condition of land underpins the whole process of managing land contamination, whether as part of redevelopment of previously developed land or for other reasons. As described in the Environment Agency's Model Procedures (see Chapter 1), the technical management process can be divided into the three main stages of:

- risk assessment
- options appraisal
- implementation

Each of these stages identifies sub-stages of the process which have particular information requirements and outputs related to the condition of the land. The overall process is one of risk management, and is aimed at systematic and appropriate identification and formulation of any problems at the beginning of the process, assessing any risks and finally leading to selection and implementation of suitable ways to deal with any unacceptable risks.

In an ideal world therefore, appropriate information on the condition of the land will be obtained and recorded at key stages to support the risk management process. At the end, information can be provided both to confirm that risks have been dealt with and to inform any subsequent actions in relation of the land, for example any further change of use.

4.2.1 *Practical issues*

In reality there are some practical issues associated with information on land condition, for example:

- **Information ranges in complexity:** information about a site ranges from relatively simple data about areas and topology used, for example to confirm layouts or calculate extent of activities, to more complex information about the geosphere used in assessing the potential interactions of the site with its environment (all of which may be recorded in a Geographical Information System – see Chapter 7). Information on the former uses of land, as indicated in Part B of this book, can range from simple chronological histories of activities on site to indicate the potential for contamination to more complex physical, chemical or even biological investigation of different chemical substances and their effects.
- **Information about land may change over time:** information is usually obtained from a variety of sources, and obtaining it may be a lengthy process. Information about land may also change as a result of actions on the land. It is essential to know the source and currency of data.
- **There will be conflicts and errors in information:** for example, as investigation techniques evolve and new types of information become available, there may be apparent or real conflicts with earlier data. The nature of site investigation is such that there are many opportunities for error in obtaining or recording information, and when that information has been processed through a number of different stages, full details and provenance of the raw data, which may be essential for the current use of the information, may have been lost.

- **Different levels of information are appropriate for different decisions:** the need for particular information on any individual site will be related to the decision stages during the process, and to some extent to the nature of the person or organisation that will be using that information. Some decisions can be made on the basis of limited information, others require a great deal of technical data about the site and its surroundings before an assessment of the implications can be made.

It may never be possible to obtain a perfect picture of any particular site. This may be either because this is impractical or because the cost involved outweighs the benefits. In practice there will always be gaps, and, even if these are not overt, all information will carry with it a degree of uncertainty. The gaps and uncertainties are not necessarily insurmountable issues, but they do need to be known so that they can be factored in as appropriate at any stage of the decision process.

4.2.2 *The need to manage information*

As a result of the practical issues, and the variability of individual sites, it can be difficult to provide absolute prescription about the amount and quality of information for every site. But inadequate, unclear or erroneous information can lead to serious problems; critical decisions made on wrong information are likely to be expensive mistakes.

The Urban Task Force (UTF) identified the benefits of greater consistency in handling information on land contamination. In their review of Urban Renaissance (Urban Task Force, 1999), focusing on the regeneration of brownfield sites, they recommended the introduction of standardised documentation describing the condition of the land. The UTF saw a particular need:

> 'To ensure that during the sale, purchase and redevelopment of land, all concerned parties had access to the same data-sets and could therefore develop some general agreement between them on the levels of risk associated with that particular site and that particular use.' (Urban Task Force, 1999, pp. 242–3)

The remainder of this chapter describes two related developments which take forward this recommendation: the Land Condition Record (LCR) and the associated Specialist in Land Condition (SiLC) registration scheme. It explains the principles behind them both, describes their role and application to the land redevelopment process, and considers how they can both help to provide greater confidence and efficiency in the reuse of previously developed land.

4.3 The Land Condition Record (LCR)

The Land Condition Record is a structured presentation and summary of available information relevant to land contamination for particular land or property. It has been developed by a working group consisting of representatives of organisations with key interests in land management and regeneration, including landowners, regulators, property and environmental professionals, developers and house-builders.

The report[1] of the working group, covering the background to the development of the LCR, its content, use and management and containing the LCR proforma, was launched in November 2000. At the launch, and in her foreword to the report, Hilary Armstrong, then Minister for Local Government and the Regions, highlighted the benefit of a robust and reliable tool for information collection and exchange in overcoming barriers to the successful commercial redevelopment of land.

4.3.1 Use of the LCR

The LCR is intended to be used in a number of areas where the collation and distillation of information on land condition into a standard format can be expected to improve the transfer of information, focus discussion constructively on any relevant gaps or inconsistencies and assist in the rapid assessment of real rather than perceived problems as part of the risk management process. Some of the specific applications can include:

- **Conveyancing:** for example the transparent provision of information on the condition of land by the vendor can be relevant to excluding the vendor subsequently from liability identified under the Part IIA contaminated land regime.
- **Development and building control:** standardised presentation of information about the site can reassure the regulatory authorities that contamination is being addressed during the process and streamline the process.

[1] A Standard Land Condition Record. Report by the Land Condition Record Working Group originating under the Urban Task Force. Available as pdf download from www.silc.org.uk.

- **Environmental regulation:** information in a systematic format can indicate whether or not there are 'pollutant linkages' likely to be present and what risk management arrangements are in place.
- **Health and Safety Management:** making collated information available regarding potential hazards that may affect those using or working on the land or property helps landowners fulfil their legal responsibilities.
- **Insurance and confidence building:** quality-assured information can reduce the risk of inadequate assessment or management of problems.

4.3.2 *Key features of the LCR*

Several issues were addressed by the working group that influenced the final shape and content of the LCR. They made four main recommendations in their report, discussed below:

> 'The LCR is to consist of available factual information or basic data relevant to land contamination.'

> This means that the LCR can be compiled from what is already known about the site, for example by bringing together data from different reports, even though this may leave gaps. The reasoning behind this is that the alternative assumption that all possible items of data must be provided for any particular land or property would result in potentially costly and unnecessary work. However, gaps in the LCR may subsequently be filled by further investigation, but only if relevant and needed by the owner or user of the LCR.

> 'The LCR "must be completed within a quality assurance system, including the use of accredited specialists in contaminated land when these become available".'

> This emphasis on quality underlay much of the thinking of the working group, and followed the main Urban Task Force concerns about the difficulties and delays that result from disagreements over data or errors in information. As a result, quality assurance information is overt within the document management section of the LCR, and the LCR format and method for compilation also follow through the requirements for data provenance and quality to be explicit. The suggestion by the working group to use accredited specialists led directly to the creation of the Specialist in Land Condition (SiLC) scheme described later in this chapter.

'The LCR "is to be voluntary and straightforward to complete and maintain".'

The working group was keen to see the preparation and transfer of LCRs becoming a matter of routine, for example to support land transactions and planning applications. However, they recognised that creating an automatic requirement for landowners or others to maintain a record of the condition of land could simply introduce a new bureaucracy. Landowners or developers therefore mainly instigate completion of the LCR. The format recommended by the working group was that of a Microsoft Word template, so that data entry could be done by those used to relatively simple electronic wordprocessing.

'The LCR "should NOT include assessments of the implications of the information, for example whether the land is suitable for use or whether pollutant linkages are present".'

The information in an LCR is structured in a way that provides for rapid identification of the possible presence of contamination and what it might affect – for example because of the use of the land or its environmental setting – and a simplified assessment of the resulting implications. However, this identification and assessment will be constrained by the particular timing and reason for the evaluation, and is therefore something that complements the information in the LCR rather than being a part of it. The working group saw the LCR as being one of a family of documents about the site.

4.3.3 Detailed contents of the LCR

The LCR is divided into the following nine main sections with a number of important annexes:

- **Executive summary:** a one or two page part of the document which summarises the key features of the site, together with any gaps or caveats relating to the information. This is designed to be something which can be provided on its own, for example to a non-technical reader of the document, but refers to the supporting full LCR information.
- **Document management information:** a quality control sheet which indicates the availability of earlier LCRs (and what changes there have been) or related LCRs (for example an LCR for a sub-zone of the site – a daughter LCR – which might contain additional information) and the details of the compiler and verifier of information

entered in the LCR. This section also indicates the status of the LCR, for example that its release is restricted because it is still in preparation.

- **Land referencing information:** a comprehensive set of data relating to the address and location of the site, together with details of local authorities and other local or regional bodies. The information ranges from land registry title numbers and OS references to a description of the location of the site, so that it can be found readily in a variety of contexts and its general setting begins to emerge.
- **Current land use and access:** a systematic identification of the current uses of the site using standard categories, with details of ownership and occupation, as well as access and provision of services. Any known potential changes to the land use are also included here. This information is particularly relevant in the UK policy context of assessing risks from contamination in relation to the use of the land, as the type of land use and accessibility of the land will indicate the relevant assessment approaches. The information on current owners and occupiers is useful for a variety of reasons, for example it may be relevant for planning any investigation work.
- **Surrounding land:** a note of the current uses of nearby land. This information may only be partial, but can indicate the setting of the land covered by the LCR and is typically obtained during the desk study stage of site investigation. It assists in preliminary assessments of potential contamination from sources off the site and of the impact of contamination on the site, for example on an adjacent Site of Special Scientific Interest.
- **Proximity to controlled waters:** not a full record of the hydrogeological status of the site (which is summarised in a later part of the LCR), but a summary of the presence of surface, ground and coastal waters which might indicate the need or otherwise for further data. This is relevant for example to the assessment of whether the site is potentially a special site under the Part IIA contaminated land regime.
- **Land history:** a systematic summary of all the known uses of the land over time, with a categorisation of the type of use (so that for example the type of use can be compared with the details of industrial processes discussed elsewhere in this book). In line with good technical practice, this section also allows for the site to be zoned, as different uses may have taken place in distinct areas of a larger parcel of land, and for details of the operational history for each use to be shown, as this can be a significant factor in identifying

the particular contamination potential. The section also presents information on what has happened to the site after the different uses, for example if an earlier use has ceased, it may be known that the site was decommissioned in a documented way.

- **Desk study and investigation information:** a summary of the visual, topographical (including ground stability and flooding), physical (including geological and structural), chemical, biological and hydrogeological information obtained from any site investigations that have been carried out on the land. The LCR does not repeat all the data in the underlying reports, but synthesises it in a transparent way so that it can be viewed in summary form. This enables the identification of any need for further data or more detailed assessment of the underlying information. Again, in line with good technical practice, the information may be presented separately for different zones of the site, and there is an opportunity to indicate any similar information that is available for adjacent land.
- **Remediation details:** an up to date picture of what has been done on the land to deal with any contamination problem. The information on remediation is summarised and categorised by typical remediation techniques, and includes details of the intended remediation objectives as well as any statutory controls and verification. The section also includes details of any ongoing risk management, which may be a critical part of ensuring that the land is suitable for use.

The annexes are an important part of the LCR. They comprise:

- **Regulatory authority information relevant to the land:** a schedule of correspondence and other formal documents, for example permits or licences, that apply to the land.
- **Record of contractual information:** a schedule of warranties, insurances, payments made for remediation, bonds, assignments of reports and other similar information.
- **Record of desk studies and investigations:** An index of studies and investigations, followed by a detailed summary for each report in a standard format giving details of the organisations who commissioned and carried out the work, what it covered and what its objectives and constraints were and what quality assurance and other information is available to support the data.
- **Other references:** a schedule of other documents which may provide data for the LCR, such as standard sources.
- **Copies of key documents:** a schedule of any documents provided with the LCR.

In many cases only part of these sections and annexes can be completed from existing information. However, this gives an overall picture of what is known about the site. It can be built up from a large number of separate reports, enabling a number of users of the LCR to work from a common summary, without the need to assess the separate reports each time.

4.3.4 *Compiling an LCR*

The report of the working group also included preliminary notes on the completion of the LCR. Following feedback on these, a more comprehensive set of guidance notes on compiling the LCR is provided on the www.silc.org.uk website. The process of compilation involves 12 key steps, which are presented as instructions to the compiler as follows:

- **Recognise purpose of LCR and note the key features:** it is intended to give a picture of the property, leaving detailed investigation reports to provide all the underpinning technical data. The LCR is to contain factual information, not speculation or implication. It uses only available information, summarising more detailed reports and giving the provenance of information.
- **Use the document template in Word software:** this was prepared as a development version for the working group by the IT specialists at Arup and is available as a download from www.silc.org.uk. It is intended that subject to funding this will be updated shortly to provide greater facility in use.
- **Locate property:** it is essential to be clear at the outset exactly which property or part of a property is to be covered by the LCR, and a map showing the location, together with a plan of the site boundaries, forms part of the LCR.
- **Assemble data:** the LCR is essentially a summary of existing information, and when first compiled will require the collation and distillation of a number of earlier reports and possibly overlapping investigations. The documents which are available are likely to include reports (e.g. desk study, site investigation), other documents such as licences, correspondence with regulatory bodies, contractual information and standard information such as OS maps.
- **Start with Annex 3:** as the underlying data are likely to be held in a number of different site investigation reports, the first step is to list these studies and reports, identify the type/parts of study (using the standard categories used in the LCR) and component

activities and then fill in summary sheets in the Annex. In doing this, the compiler will establish a picture of the data that are available, and become aware of overlaps and contradictions, as well as the quality of underlying data.

- **Complete main LCR starting at Section A2:** leaving the summary until last, the compiler should go through all sections, focusing on the information specified in the form. All boxes are to be completed or indicated as having no information (see next step), although the form has some check boxes for fast track where substantial sections can be left blank. All information is to be referenced to one or more of the documents indexed in the different annexes. For example the information that a particular former use has been identified on the land may be referenced either to a desk study report (which in turn may refer to a primary source) or to the primary source (such as an OS map) if this has been made available, for example as one of the documents listed in annex 4. The intention is that any information given in the LCR can be traced to its source.

  The working group identified three circumstances where no information is available: *not known* (i.e. where there are simply gaps), *not found* (for example where mining records were examined but no evidence of mining was found), *not examined* (for example where a document was not released to the compiler). Each of these has different implications and the phrases are used deliberately in the compilation of the LCR.
- **Identify and deal with contradictory data:** this is an inevitable feature of land contamination information, for the reasons discussed earlier in this chapter.
- **Consider surrounding land:** in many cases the LCR will concentrate only on a particular property. However, information may be available about surrounding property and the LCR provides a number of opportunities to record this, indicating which land has been considered. (For example, it is typical to have an indication of certain types of use of land within a 1 km radius of the site.)
- **Review gaps, caveats and assumptions:** these are all highly relevant to the transparent and efficient presentation of information. At certain points in the LCR there are prompts to indicate these as specifically as possible. Gaps are where information is not known or not examined; caveats are where the compiler has reservations about the quality of data and assumptions may have been indicated in the underlying reports or have been used by the compiler, e.g. where they have dealt with apparent typographical errors in underlying data.

- **Check references and dates:** the LCR prompts the compiler to give the references for the information entered. In some cases this is a general reference to the report listed in a schedule (for example 'Report reference: A3/01'). In others it is a reference to a specific part of a report (for example 'Ref in report: A3/01 p. 11').
- **Prepare summary:** this is to be a stand-alone text, summarising key features of the LCR. It should cover all sections of the LCR to provide an overall picture of what is known.
- **Issue LCR within appropriate document management system:** at this point any disclaimers or reservations about the use of information should be indicated in the standard front section.

Finally, those compiling the LCR are likely to have formed a professional view on the significance of what is known and what is not known, and, if this is necessary, how to fill gaps. This can form part of the normal process of professional advice to clients or management of a site within an organisation. As further information is obtained the LCR can be updated, with successive versions indicating changes.

4.3.5 *Future development of LCRs*

The description above indicates the scope and approach taken in developing the LCR format from the working group recommendations. As the LCR has been used over the last few years further improvements, particularly to the usability of the electronic version and to the increased standardisation and simplification of information within the LCR, are emerging.

Although relatively few LCRs have been compiled for sites – reflecting amongst other things the inevitable costs and step change in attitude required – there is increasing interest in the benefits. One of the significant features of the last few years has been the interest from a number of local authorities in receiving information on a site in the form of an LCR to improve their ability to assess quickly the condition of the land to fulfil their responsibilities in the planning and development control process.

Other local authorities have used LCRs to pull together information about sites being considered under the Part IIA process, especially where there is a complex history of multiple reports. The collation of the information has helped to pinpoint issues to resolve, although in some cases it has also helped to identify where there is a relatively simple explanation of the apparent duplication and inconsistency between site investigations.

A number of landowners and developers are increasingly using the LCR to provide the full history of the information they have on the site, and what has been done to remediate it. LCRs prepared by registered SiLCs are increasingly being seen as a way of providing simultaneously greater reassurance to the regulator and a transparent alternative to regulatory sign off.

4.4 The Specialist in Land Condition (SiLC) Registration scheme

As indicated above, the working group represented a wide range of interests in the property and environment sector. They felt strongly that confidence in information about land condition would be improved by the creation of a register of appropriate experts, a suggestion which had been mooted on a number of previous occasions. Although the immediate need was to ensure that LCRs were signed off by registered specialists, the general, and increasingly apparent wish, was for the experts to be clearly identifiable as those who could also be relied on to evaluate and make recommendations about land condition.

As a result, a number of individual members of the working group were encouraged to initiate the development of a registration scheme for specialists, working with their professional institutions. At the outset it was recognised that not only was there a need to create a register that recognised the multi-disciplinary nature of land condition issues, but also to create a formal scheme supported by the chartered institutions and other senior bodies representing the different underpinning professions.

As with any scheme of this nature, the credibility of the register also relies on the transparency, independence and quality of the selection process, and ultimately will depend on the performance of those registered.

4.4.1 Management of the scheme

With the support of a number of key institutions, a Professional and Technical Panel (PTP) is responsible for developing and overseeing the registration scheme. The panel includes both technical and professional administrative members, with representatives from:

- Association of Geotechnical and Geoenvironmental Specialists
- Chartered Institute of Environmental Health

- Chartered Institution of Water and Environmental Management
- Geological Society
- Institute of Environmental Management and Assessment
- Institution of Civil Engineers
- Royal Institution of Chartered Surveyors
- Royal Society of Chemistry

The Institution of Environmental Management and Assessment (IEMA) provides the secretariat for the panel and administers the scheme. IEMA provides a dedicated website for the scheme (www.silc.org.uk), runs training courses, and coordinates publicity about the scheme. Funding to get the scheme off the ground has been provided by the Regional Development Agencies, through a South East England Development Agency (SEEDA) grant for 3 years.

4.4.2 *The requirements for a SiLC*

The PTP agreed that the minimum eligibility requirements for becoming a SiLC would be 8 years of suitable experience in addition to membership at an appropriate level of a professional institution recognised by the scheme. In addition, the criteria for becoming a SiLC is that a candidate should be able to:

- Show familiarity and understanding of all sections of the Land Condition Record (LCR).
- Describe the contents and completion of each section of the LCR.
- Describe the uses of the Land Condition Record; this should include those set out within [the report of the working party on the LCR].
- Interpret information contained in the LCR and give advice to landowners and other suitable parties based on this interpretation.
- Demonstrate a thorough knowledge of their field in relation to the LCR.
- Demonstrate awareness and some understanding of other relevant fields and connected professions.
- Demonstrate objective judgement in information and data management.
- Communicate well.
- Manage effective interaction between clients and other interested parties.
- Know and demonstrate a willingness to comply with all sections of the Code of Practice.

The ability of a candidate to meet these requirements is tested over a three-stage process. The initial process included an open book exam with technical and other multidisciplinary questions as the second stage, marked blind by assessors drawn initially from the panel and then from successful candidates. This scheme was piloted with a number of nominated candidates, and subsequently ran for a number of rounds. The result is that a pool of some 46 SiLCs now exists (December 2003) to provide peer assessors. The scheme is to be modified in 2004 to reflect this, streamline administration and provide more flexibility for candidates to fit the application into the demands of their everyday work.

The revised process also comprises three stages:

- **The application:** where the eligibility of the candidate is verified by IEMA, if necessary by referral to the panel.
- **A written paper:** which includes questions related to the LCR (e.g. completion of an LCR based on sample data and a follow-up question testing the candidate's understanding of the resulting LCR), together with an individual written submission demonstrating how the candidate meets the remaining criteria for becoming a SiLC.
- **A peer interview:** following assessment of the written paper, two assessors probe the candidate to confirm that they meet the SiLC criteria. Throughout the operation of the scheme the assessors have been asked to declare any conflict of interest – for example professional relationships or contact with the candidate – which precludes them from interviewing. A representative from IEMA attends the interview to ensure that the appropriate process is followed.

If successful the individuals are identified on the register maintained by IEMA as Specialists in Land Condition, and allowed to use the letters 'SiLC' with their professional qualifications. Each SiLC has a personal registration stamp, such as that in Fig. 4.1. SiLCs are required

Fig. 4.1 The SiLC logo. Reproduced with permission.

to maintain their continued professional development (CPD) in relation to land condition issues and provide evidence of that to IEMA to remain as a SiLC. The PTP reserve the right to re-assess any registrant after 5 years.

4.4.3 *SiLC Code of Practice*

The Code of Practice, referred to above, requires that a Specialist in Land Condition shall:

- uphold and promote the integrity of their profession
- exercise honesty, diligence and impartiality in their professional work
- seek to understand and comply with all legislation/standards in the country in which they are practising
- not allow conflict of interest to influence their professional decisions and judgement, make all bodies concerned aware of such conflicts
- not accept anything of value from clients, employers or third parties which could be thought to influence their professional judgement
- continually work to maintain and improve their knowledge using a combination of training and practical work; and give reasonable assistance to candidates wishing to enter the profession
- maintain and enhance levels of proficiency, both individually and throughout the profession
- when giving advice, make suitable persons aware of the potential consequences and alternatives
- acknowledge their limitations of competence and not undertake any work which he/she knows is beyond their professional capability
- ensure all information given to and contained within the land condition record is, to the best of their knowledge, correct and accurate
- not endorse any information or declarations from clients or third parties which they cannot verify as accurate and true
- endeavour to uphold and enhance the reputation of the register

4.5 Conclusions

The importance of maintaining an accurate, and up-to-date, record of land condition cannot be emphasised too strongly. Information pertaining to land usage and its physical condition may be required for many purposes, including raising finance for business purposes, preparing redevelopment proposals and in satisfying regulators with regard to suspected contamination. Therefore records relating to land condition should be updated every time there is a change in site activities and at the time of every transaction involving a sale or lease of the land.

The two initiatives of the Land Condition Record (LCR) and the Specialist in Land Condition (SiLC) register are intended to improve the quality and confidence in information about land and the evaluation of that information. They are an increasingly vital part of the process – described in this book – of assessing and managing the varied issues related to the regeneration of previously developed land.

Checklist

- Is information available relating to the condition of the land and is it in a format that is readily accessible?
- When was the information last updated and have there been any changes in site activities, or any disposals, since the last updating?
- Who prepared the reports and/or record and for what purpose?
- Is any additional information required?

Chapter 5
A Few Legal Predictions

Paul Sheridan

5.1 Introduction

It is often said that environment issues are largely driven by regulatory law and policy. Whilst this is not completely correct it is a workable premise for present purposes.

The first phase of modern environment laws was largely directed at dealing with, and preventing, pollution and contamination. Subsequent phases of environment law have taken on a much more preventative focus with an emphasis on internalisation of the costs of the 'environment externalities', namely that (where possible) the broader effects on the environment of products, services and operations should be internalised into the initial costs of those products, services and operations. Whilst these newer laws are only now starting to bite in the re-use and redevelopment of previously developed or 'brownfield' land, the stakeholders in such developments will need to look not only at pollution and contamination laws but also environment laws impacting upon the supply chain and the performance of the development once completed.

Environment law and related legal processes have to date often been regarded as impediments to development. It may be argued that environment law gives rise to delay, extra cost and uncertainty. Certainly some aspects of some environment laws are difficult to interpret and can thus give rise to uncertainties, leading to further delays whilst the various parties engaged in the redevelopment process endeavour to reach consensus. Further, the interrelationship between some environment laws (and notably those relating to contaminated land and waste management) is unnecessarily complex. Current implementation and enforcement of environment laws and policies is also highly criticised for its lack of consistency and the slowness of related decision making.

In relative terms environment law is not a mature body of law but instead is a body of law that is suffering from growing pains. One of the most significant problems is the lack of access at speed and minimal cost to decision makers or tribunals for clarification of the true meaning of disputed aspects of the laws and the application of those laws to particular circumstances. When this issue is coupled with the piecemeal implementation of many environment laws and the low enforcement rates, it is easy to see how criticisms arise.

In spite of these criticisms environment laws are beginning to provide opportunities for the redevelopment of previously developed sites, including those affected by contamination. Environment law is giving rise to a healthy environment industries sector, which is growing exponentially. This sector clearly requires land and more often than not the sector's land requirements can be found on brownfield sites, and often on the more difficult of these sites. Further, the industry is creating remedial techniques that ultimately will reduce the volume of construction waste that currently is disposed of at landfills thereby effecting potential savings of considerable costs in the light of the effects of the Landfill Directive (see below).

With the growth and complexity of environment laws applicable to brownfield developments, the exposure of professional advisers correspondingly increases. Environment law provides significant risks for designers, project managers, consultants, surveyors, lawyers and the like. These risks have to be matched by suitable risk management measures.

5.2 Relevant laws

Below are comments on some (but by no means all) of the environment laws that are relevant to the redevelopment of previously developed and, especially, contaminated land. The purpose of this section is not to delve into the detail of these laws. Instead it is simply to draw out some salient points. A general point that can be made in relation to all of these laws is that they each contain problems of interpretation and problems relating to the interrelationships between each other.

5.2.1 Statutory Contaminated Land Regime – Part IIA Environmental Protection Act 1990

The Environment Agency has estimated that there are approximately 100 000 sites in England and Wales that may be affected by contamination, of which 5–20% may require remedial action. As noted in Chapter 2 the number of sites to date determined as 'contaminated land' or 'special sites' is fairly small. However, care needs to be taken in reading the statistics in this area, because one of the main policies behind the statutory contaminated land regime was to encourage voluntary remediation through transactions and developments, rather than obligatory remediation through the statutory regime. Voluntary remediation is certainly occurring on a much greater scale than that required by regulatory action.

The Statutory Contaminated Land Regime is complex, with difficulties that are both legal and practical in nature. Many of the legal difficulties remain to be answered at law. However, one of the hurdles has been overcome: namely, that the threshold for water pollution in the second part of the definition of contaminated land has been amended so that only 'significant' pollution of controlled waters or the 'significant possibility of significant pollution of controlled waters' will trigger this legal regime. This change was brought about by the Water Act 2003.

A practical problem faced by those who seek to carry out on-site remediation (whether in situ or ex situ) is whether the laws on waste apply to this remediation. The conceptual argument is that the contaminants are waste and therefore waste law applies. This can cause significant problems in practice. In October 2002, a working party (made up of representatives from various industry sectors and the regulators) set up under the auspices of the Urban Task Force produced a report 'The Remediation Permit – Towards a Single Regeneration Licence'. This is being taken forward as an integral part of DEFRA's current review of waste permitting and an interim report was submitted to Ministers in July 2003 (DEFRA, 2003a). For further information see the waste permitting review at www.defra.gov.uk/environment/waste/legislation/permitreview/index.htm.

The EU's proposed Directive on Environmental Liability with regard to the Prevention and Remedying of Environmental Damage is currently going through the legislative process at the EU. In the UK DEFRA undertook a consultation process on the Directive in April and May 2003. Of the 190 organisations consulted by the Department only 27% (51) submitted responses. Of the respondents, over 80% supported the broad principles of the proposed Directive whilst a minority (5%) felt that the proposal did not go far enough. Approximately three quarters of those supporting the broad principles did not think the proposed Directive would achieve its environmental aims, citing the need for greater clarity, particularly in the definitions (see www.defra.gov.uk/environment/consult/liability2/response/index.htm for further details about the consultation).

Once adopted, consideration will need to be given to whether any significant changes will be needed to the UK's domestic laws on contamination and pollution prevention, in particular the Statutory Contaminated Land Regime and the Works Notices Regime under Section 161 A-D of the Water Resources Act 1991.

5.2.2 *Part C of the Building Regulations*

Part C of the Building Regulations, which deals with Site Preparation and Resistance to Moisture, is to be amended. A consultation process was started in December 2002 and it is proposed that the amendments to this part of the Regulations will deal with issues relating to contamination, climate change and flooding.

The proposed changes relating to contamination will bring the Regulations up to date with the concepts underlying the Statutory Contaminated Land Regime. However, three significant changes are proposed to the current Part C of the Building Regulations. The first is that the whole of the site, and not just the footprint of the building, will come within the Regulations. The second is that the amended Regulations will apply to material changes of use where future residential use or other form of accommodation is proposed. This could prove important in the light of the current emphasis upon mixed-use redevelopment schemes. The third is that these amended regulations conceivably could introduce a potential 'sign off' for contaminated land remediation works. Under the Statutory Contaminated Land Regime there is no official sign off for remediation works. However, under building regulations a certificate of completion is normally obtained and the query must be, therefore, would this certificate of completion evidence a sign off for the purposes of the Statutory Contaminated Land Regime? This is probably not the intent but we shall have to wait and see whether the completion certificate can be used to this effect.

The Council for British Archaeology was concerned to find that it was not on the list of organisations consulted by the ODPM and noted that it could 'find no mention of the need to take accoun[t] of archaeological deposits or historic and listed buildings in these proposals. Decontamination and flood prevention measures can have major implications in terms of loss of irreplaceable fabric and/or deposits if their impacts are not properly assessed and planned for' (Council for British Archaeology, 2003).

5.2.3 *Asbestos*

Much has been written about the Control of Asbestos at Work Regulations 2002, which were brought into force in October 2002. They bring about important changes. Clearly, many brownfield sites contain asbestos and, therefore, these Regulations are important. The Regulations impose obligations to actively identify, record and manage asbestos. This duty came into effect from 21 May 2004. It is imposed upon anyone with responsibility under a contract or tenancy for the maintenance or repair of non-domestic premises or where there is no such contract or tenancy the person who has control over the premises.

Not surprisingly, the existence of asbestos is, or should be, a significant due diligence issue. It should also be a significant issue in the drafting of construction contracts and contracts involving PFI/PPPs. Interestingly, due to concessions by the Inland Revenue it may be possible for a corporation with an interest in the land to set off against corporation tax much of the costs of the asbestos work at the enhanced rate of 150% under the Contaminated Land Remediation Expenditure Relief.

5.2.4 *Water Framework Directive (2000/60/EC)*

This Directive is potentially a sleeping giant in a number of respects. With regard to brownfield land, the Directive will ultimately impact upon the planning regime by forcing local authorities to pay closer attention to the water environment. The Directive is complex and is riddled with legal difficulties. It requires Member States to do various things to ensure that surface water reaches what is called 'good ecological and chemical status' by 2015 and that groundwater reaches what is called 'good quantitative and chemical status' by the same date. Certain priority substances and priority hazardous substances are to be phased out or banned. Importantly, water resources are to be managed on a River Basin District Basis with management plans to be drawn up for each district. In England this will be coupled with the CAMS (Catchment Abstraction Management Strategies), all of which will ultimately feed into the planning process. Further, water use, and in particular water abstraction, is to be changed to take into account the full cost of its environmental impacts (this is an example of the internalisation of environment externalities as referred to previously).

5.2.5 *Japanese knotweed*

The existence of Japanese knotweed (*Fallopia japonica*) has recently become a due diligence issue, particularly in relation to brownfield site developments, that can impact upon the professional work of property consultants, building surveyors, lawyers, architects, environment consultants and project managers. Japanese knotweed is a pervasive plant that can cause harm to indigenous species. It is an offence under the Wildlife and Countryside Act 1981 to cause Japanese knotweed to grow in the wild, although it is commonly found along railway lines, riverbanks, roads and footpaths, in graveyards and on derelict sites.

Leaving aside the potential criminal offence, within the context of a development Japanese knotweed can cause significant logistical and financial issues because of its capability of growing through or damaging loose paving, tarmac and sub-surface services and because of the potential impact on value. For obvious reasons developers are keen to ensure this plant's destruction or otherwise to effectively restrain its destructive power. Japanese knotweed can be dealt with by means of herbicides but this can often take two to three growing seasons to achieve. Obviously, such time is rarely available within a development. Often physical protection measures are required and these can be costly. The Environment Agency has published a 'Code of Practice on the Management, Destruction and Disposal of Japanese knotweed' (www.environment-agency.gov.uk/commondata/105385/knotweed.pdf). This is a useful background document but more considered advice is usually required within the context of a redevelopment. Unfortunately consistent professional advice in this regard is often difficult to obtain. It must also be borne in mind that the waste created from dealing with Japanese knotweed must be dealt with in accordance with waste law.

Japanese knotweed is often found on the boundary between different parcels of land. Clearly, a rudimentary risk assessment will identify that removing Japanese knotweed from only one of the parcels of land will probably not be adequate as it could simply grow back from the neighbouring land. Consulting or negotiating with the neighbour and trying to deal with the Japanese knotweed on both parcels of land is often required.

5.2.6 *Waste Law*

The interrelationship between waste law and on-site remediation has been mentioned previously. Otherwise, there is a simple but powerful message being given by the UK's developing waste law: our waste culture is in for a shock. The Landfill Directive (99/31/EC) will have significant impacts upon our waste culture. The landfilling of waste is to be substantially reduced. Hazardous waste landfills could become very scarce. Certainly the cost of waste to landfill will increase and it may be that haulage distances to landfills will be greater (and hence haulage costs will be greater). Project managers and developers will want to understand and address the cost and logistical implications of these changes.

There are a number of directives in waste law which are now aimed at what is called 'producer responsibility'. In short this means that the producer of a product must internalise into the cost of the product the ultimate waste management cost of the product at the end of the product's life. For instance, the Waste from Electrical and Electronic Equipment Directive (2002/96/EC) requires Member States to set targets for the recovery and recycling of waste from electrical and electronic equipment and to minimise the amount of such waste that goes to landfill. The producers of the equipment are to internalise the waste management cost into the price of the product. Thus again project managers and developers will want to understand the potential increases in the costs of equipment used in the fitting out of buildings, such as ovens, fridges, lighting, etc.

5.3 Economic and fiscal instruments

5.3.1 Contaminated Land Remediation Expenditure Relief

This tax relief was implemented by section 70 and Schedule 22 of the Finance Act 2001 and is now fairly well known. The measure is intended to 'make the development of [contaminated] sites more viable, helping to tackle the legacy of previous industrial uses and reduce the pressure to develop greenfield sites' (HM Treasury, 2001, para 6.82).

The tax relief is available in respect of any estate, interest or right in land in the UK acquired by a company (i.e. the relief is not available to legal entities other than companies). The relief applies to land that is contaminated but not where the contamination occurred through an act or omission by the company (i.e. the tax relief does not apply where the company is the polluter).

Where the company incurs certain capital and revenue costs in cleaning up the contamination the company may deduct that expenditure, at the enhanced rate of 150%, in the computation for tax purposes of the profits of a Schedule A business or a trade carried on by the company (there are also fairly usual provisions for tax credits in the event that the company is making a loss rather than a profit). The relief applies to expenditure incurred after 1 April 2001. The relief will not be available to the extent that the expenditure is subsidised by any grant or subsidy or is otherwise met directly or indirectly from a third party (for instance under an indemnity from the vendor of the land).

The relief has proved very useful particularly as it may extend to cover asbestos works. According to advice given at the Large Corporate Forum in November 2002, it is the view of the Inland Revenue that the relief does not extend to the costs of dealing with the presence of Japanese knotweed (see www.inlandrevenue.gov.uk/lbo/minutes_201102.htm). The tax relief is to be reviewed after five years of operation (Inland Revenue website, www.inlandrevenue.gov.uk/budget2001/revbn22.htm).

5.3.2 *Landfill Tax*

Landfill tax is paid on the volume of waste deposited at a landfill. Inert wastes attract a tax of £2 per tonne. The rate of tax for non-inert waste in the tax year 2003/2004 was £14 per tonne, rising to £15 per tonne in tax year 2004/2005. The Government's current proposals are that this landfill tax should rise by an accelerator of £3 per tonne from tax year 2005/2006 until such time that it reaches £35 per tonne (HM Treasury, 2003, para 7.46). It is to be noted there are many who argue that this accelerator is not steep enough and that £35 per tonne does not truly reflect the cost of the related environmental damage. It is also to be noted that with the proposed introduction of a truck road user charge (see below) the cost of haulage of waste offsite may increase.

There is a statutory exemption from landfill tax for certain contaminated soils. Care must be taken in the planning of remedial works in order to obtain the benefit of this exemption.

5.3.3 *Proposed Truck Road User Charge*

Announced in the 2002 budget, this is another example of a fiscal instrument aimed at internalising the costs of external damage to the environment. In that budget the Government stated that it was committed to helping the haulage industry become more competitive and less damaging to the environment. The Government stated that it 'believes that all lorry operators using UK roads should make a fair contribution towards the costs they impose, irrespective of nationality. As announced in Budget 2002, and subject to the outcome of discussions with potential suppliers, the Government aims to introduce, by 2006, a road-user charge based on distance travelled'. Interestingly, however, the Government also stated that it wished to offset this charge against other related tax. Following further studies the Government came to the view that the charge should be offset by a reduction in fuel duty. At face value it appears inconsistent to impose a charge aimed at benefiting the environment and at the same time offsetting that charge against fuel duty, a duty that is also said to be imposed to benefit the environment. Many have suggested that this offset is in fact aimed at securing a charge on non-UK trucks that use the roads of the UK having first filled their tanks with cheaper diesel purchased on the European continent and thus avoided having to pay fuel duty to the UK exchequer.

5.3.4 *Aggregates Tax*

The Aggregates Levy was introduced in April 2002. The aim of the tax is to reduce demand for virgin aggregates and encourage the use of recycled materials, such as crushed concrete from previously developed sites, as well as to address the costs to the environment associated with quarrying, e.g. noise, dust and visual intrusion. The tax applies to sand, gravel and crushed rock and is charged at £1.60 per tonne. The levy will not apply to coal, clay, metals, gemstones and industrial minerals.

5.3.5 *VED differential rates*

The Government is committed to reducing emissions of carbon dioxide (CO_2) from cars and improving local air quality. CO_2 is one of the six greenhouse gases which contribute to climate change. The system of VED based on CO_2 sends a clear signal to vehicle manufacturers and purchasers about the impact on the environment of the cars they make and use, and encourages the use of more fuel-efficient cars (DVLA website, www.dvla.gov.uk/gved). VED differential rates are applicable to cars first registered and licensed on or after 1 March 2001. Prior to first registration, new cars are allocated a CO_2 figure (shown in g/km) measured as part of the car's Type Approval Test; this is fixed for the life of the car, provided that the fuel type is not changed.

5.3.6 *Enhanced capital allowances*

One hundred per cent of the capital cost of purchasing and installing certain environment-friendly equipment (such as boilers and water-efficient technologies) can be set off against turnover in the year of expenditure. The details of these allowances and the equipment involved are set out in sections 45A–J and 46 of the Capital Allowances Act 2001 (as amended). Anecdotally some customers stated that their designers do not make them aware of this capital allowance whereas some designers say that the customer is simply not interested. In any event, this is effectively 'free money' and the take-up is beginning to increase.

5.3.7 *Dereliction Aid*

Proposals to free up brownfield land and ease pressure on the green belt were approved by the European Commission in July 2003. The state aid approval aims to help the private sector transform contaminated or brownfield land into clean sites ready for development.

Announcing the European Commission's approval, the Deputy Prime Minister, John Prescott MP, said:

> 'This approval is great news for the delivery of my £22 billion sustainable communities plan. This means that sites now lying derelict because they are too expensive to clean up and unsuitable for development can be brought back into use.
>
> Areas of contaminated or derelict land ruled out in the past due to unprofitable overheads will now become more attractive to developers – reviving business in otherwise economically declining towns and villages.
>
> By bringing brownfield land back into use, this approval will take pressure off our valuable green belts.
>
> This decision will affect areas all over the country from the Thames Gateway in the South to former coal mining areas in the North and Midlands.' (ODPM, 2003c)

The approval allows the Regional Development Agencies, English Partnerships and local authorities to provide grants of up to 100% to developers to fund the cost of remediating land and tackling dereliction in otherwise economically unviable areas. Dereliction aid allows the provision of support in respect of any land, building structures or works which, by virtue of their former use and state of disrepair, cannot be made suitable for any new use without incurring costs, to remove the damage caused by previous use where this would otherwise not take place due to market failure. The approval also allows for grants to be given to help the relocation of companies on environmental grounds. The basis for calculation of the aid is the net cost of carrying out the works, after deducting any enhancement in the value of the land. For further information see www.odpm.gov.uk.

5.4 Energy and climate change

A central aspect of policies directed at combating climate change is the consumption of energy generated from carbon sources. Energy use in buildings is likely to become a significant design issue for future new builds and major renovations of existing buildings. Conceivably, it will also become a due diligence issue (particularly for institutional investors and perhaps banks) on the buy/sell or leasing of buildings.

5.4.1 *Part L of the Building Regulations*

The Building (Amendment) Regulations, SI 2001/3335 and the publication of the 2002 editions of Approved Documents L1 and L2 brought in major improvements in energy performance standards in the Building Regulations. However, these changes were only the first of four stages of proposed improvements in the requirements and associated approved guidance that are expected to be introduced in the period up to around 2008. Part L is to be again revised in 2005 and the likelihood is that standards will only increase. In October 2003 the ODPM published a paper as an initial informal response on the sorts of improvements that could be achievable in the amendment now proposed for 2005 (ODPM, 2003d).

5.4.2 *Energy Efficiency of Buildings Directive (2002/91/EC)*

This Directive which was passed by the EU in December 2002 is to be implemented by Member States by January 2006. In the United Kingdom this will probably be done through amendments to Part L of the Building Regulations and Planning Guidance.

This Directive is a central plank of the EU's climate change programme. This is because (whilst estimates vary) it has been claimed that between 40–50% of energy consumption (and hence indirectly 40–50% of emissions of carbon dioxide) is by non-industrial buildings. The Directive will require that new buildings (i.e. where construction starts after 4 January 2006) and major renovations (if the useful floor area is over 1000 m^2) will require higher energy performance standards and in some instances will need to incorporate alternative sources of (green) energy.

5.5 Corporate governance and financial reporting

If only by virtue of the volume of laws that attach to brownfield sites, never mind the risks associated with the relevant environment laws, corporate governance and financial reporting are likely to become increasingly relevant to corporations involved in the development of brownfield sites. Public limited companies involved in brownfield sites will be well aware of the terms of the Combined Code (recently updated in July 2003). Principle C2 of the Combined Code states that 'The board should maintain a sound system of internal control to safeguard shareholders' investment in the company's assets'. It also states:

> 'The board should, at least annually, conduct a review of the effectiveness of the group's system of internal controls and should report to shareholders that they have done so. The review should cover all material controls, including financial, operational and compliance controls and risk management systems. Companies which do not have an internal audit function should from time to time review the need for one' (Financial Services Authority, 2003).

In July 2002 the Government issued its White Paper: Modernising Company Law. Two points discussed in that White Paper are relevant here. The first is that the Government states that in exercising their duty to promote the success of the company, directors should have regard for the company's impact on the environment 'which the Government believes every director needs to consider as first among equals' (DTI, 2002). Whilst it is not clear what 'first among equals' means, by making this statement the Government is sending a message that a company's impacts on the environment is an issue for directors to consider and is not simply an issue for the regulators or society at large.

The second interesting point that arose out of the White Paper is in regard to economically significant companies (which probably excludes SMEs), there may be a new 'operating and financial review', rather than simply financial statements. In that operating and financial review the directors are to turn their minds to whether information is required on the company's policies on environmental issues and its performance in carrying out those policies. The Government has stated that a step change will be required in relation to more qualitative forward looking reporting rather than just quantitative historic reporting.

Financial Reporting Standard 12 requires that companies disclose and make provisions in respect of certain contingent liabilities. The most often quoted examples of these liabilities are environment risks. Certainly it is conceivable that soil and groundwater remediation costs or asbestos costs are of a nature that could require such disclosure and provisioning.

5.6 Conclusion

The number, complexity and variety of laws and legal instruments that are potentially applicable to the re-use and redevelopment of previously developed land have increased significantly over recent years. Some of these will cause increased costs, delay and uncertainty, whilst others will encourage and stimulate reuse or redevelopment. Either way extra management time will be incurred. Many of the laws will require behavioural changes. In particular, greater thought by management will be required in connection with the wider costs to the environment brought about by the redevelopment or re-use of the land and how (or by whom) those costs will be met.

Checklist

- Does waste law apply to the chosen method of remediation?
- How do amendments to the Building Regulations affect the proposed development?
- If asbestos is likely to be present, take care in the drafting of contracts and check whether tax relief will be available.
- Landfill Tax liability should be considered and, where no exemptions apply, should be factored into development costs. Steep increases in this tax are proposed.
- Government intends that companies will become increasingly responsible for the cost of recycling products, particularly electrical, in order that fewer such products end up in landfills.
- Company directors will be required to have regard for the company's impact on the environment and they may need to publish an operating or financial review.

Chapter 6
Modernising the British Planning System

Ted Kitchen

6.1 Introduction

The key elements of the British planning system have been in place since they were constructed via the Town and Country Planning Act 1947. That Act did not merely assemble a system which in its essentials has now lasted for in excess of 50 years, it also made some important assumptions about what people were entitled to expect from the operation of the system, and about the nature of the world in which those operations would be taking place (Kitchen, 2002). This chapter argues that it is profound changes in these latter elements, just as much as the inherent desirability of re-examining the performance of a planning system which is increasingly showing its age, that have led to the need to 'modernise' that system in order to make it fit for purpose for the world of the early years of the 21st century.

The chapter first establishes what the key component elements of the British planning system[1] actually are. It then looks at how some of these have fared in the face of extensive changes in the world of which they are part, over a 50-year period. Third, it summarises what the current Government is saying about the case for modernisation, before looking at the main elements of the proposed changes.[2] Alongside significant systemic changes, the Government is also proposing to address the need to change 'the culture of planning', and so the

[1] Historically, it would have been reasonable to have talked of this as a 'British' system, since differences between England, Scotland and Wales (but not Northern Ireland) were not very significant. However, in more recent times these differences have become more marked, as this chapter discusses. So when discussing recent change proposals the chapter is actually referring to the system in England.
[2] These were proposals at the time this chapter was drafted but, by the date of publication of this book, they had been enacted via the Planning and Compulsory Purchase Act, 2004.

chapter explores what this may involve. Finally, it draws together some of the key challenges to which the proposed changes give rise.

6.2 The component elements of the British planning system

The system established by the 1947 Act was in its essential elements a national system that was locally administered. In simple terms, its primary purpose was to make places better for people through a process of directing and controlling development on the part of the state. The state, via one or more of its many arms, might also be the body commissioning or carrying out development; but its most fundamental roles were first to provide a framework (a plan) to guide future development and then, second, to sanction or otherwise all but the smallest individual development proposals. There were two major reasons for instituting a system of this nature. First, it was seen as preferable to more laissez-faire alternatives, either the absence of a public role throughout most of the recorded history of urban development or its relatively toothless predecessors throughout much of the first half of the 20th century. Second, it was part of a postwar settlement which re-balanced the relationship between the state and its citizens, in order to provide for the latter a better quality of life than had been achieved previously by a less proactive state (Briggs, 1983, pp. 288, 289). Planned localities were seen as having more to offer to their inhabitants than a continuation of the development track record achieved without an effective planning system, and were seen as part of what citizens had the right to expect from the new concordat between the state and its people (see Cherry, 1988, Chapters 5 and 6).

The major planning tools established by the 1947 Act were the development plan as the forward-planning instrument, the development control process as the means of testing public acceptability of development proposals, and a series of what might be described as environmental powers, including the power under certain circumstances to acquire land and buildings compulsorily. Although there have been many detailed changes to these elements in the intervening years, in essence they remain the key tools with which the planning system has to work. Indeed, in recent years successive governments have taken steps to reinforce the significance of the development plan as the main driver of the planning process (DoE, 1997, paras 39–46), with the present government's proposals (DTLR, 2001, para 4.2) stressing the importance of a plan-led system of development control.

They also emphasise the need to engage the public more effectively with the process of plan-preparation if this role in relation to development control is to be effective. This seems to be very necessary, in the face of some evidence that public involvement in development plans processes (McCarthy *et al.*, 1995, Chapter 2) may not be as significant numerically over the time-period of plan preparation as is public involvement with the immediacy of development control consultations over a similar number of years (Kitchen, 1997, p. 82). Thus, there are undoubtedly issues around this relationship between development plans, development control processes, and public engagement which are at the heart of debates both in planning theory (Albrechts, 2002; Brooks, 2002, Chapter 9) and in planning practice (Whitney, 2003), but if anything the recognition of this as a central tension reinforces the point about the continuing role played by these major tools.

In terms of the basic structure of the system, the central role for most of the period since the passage of the 1947 Act has been played by local planning authorities (Cherry, 1988, Chapter 6). This reflected the importance of the local element in the concept of 'making places better for people', since 'place' was seen as being mainly about issues best understood from the ground upwards rather than from Central Government's national perspectives downwards. From the 1990s onwards, however, the role of Central Government in the system appears to have become of increasing importance, as more national policies have been developed and more guidance has been issued (Tewdwr-Jones, 2002, Chapter 5), to the point where it could be argued that local discretion has become unnecessarily fettered. In addition to this duality, a third element has been the scope for and the nature of a regional role in the planning process, which has waxed and waned over the years (Wannop, 1995) but which at the time of writing is seen by the Government as a key component of the future planning system (DTLR, 2001, paras 6.24–6.26). This too will be seen in some quarters as an element that fetters local discretion, although equally it could be seen as providing a helpful context for the many issues that do not respect local authority boundaries. This very brief history shows that the relative significance of these three tiers of planning activity has changed over time and continues to change, with inevitable consequences for the work of local planning authorities as the most localised of these three tiers.

In broad terms, this section of the chapter has been about the enduring component elements of the British planning system over a period of in excess of 50 years. The next section looks at some significant change elements over this period.

6.3 The passage of time

Changes in the second half of the 20th century that impacted upon the operation of the British planning system were many and varied, and it is not possible to do full justice to all of them, although their significance can readily be understood from a brief listing of some of the most important ones:

- In 1947 the most pressing urban need was to reconstruct bomb-damaged areas in Britain's major towns and cities. This need undoubtedly had an impact on some of the powers and duties written into the new Act. The primary challenges faced by the planning system today in seeking to manage urban development are very different to this, although occasionally there still remains a need to tackle something very similar such as the issues raised by the IRA bomb in Manchester city centre in June 1996 (Kitchen, 2001) (see Figs 6.1 and 6.2).

Fig. 6.1 Manchester after the bomb, June 1996. The Arndale Centre is on the left of the figure. The office building on the right and Marks and Spencer store were so badly damaged that they had to be demolished.

Fig. 6.2 The new Marks and Spencer and Selfridges building, on the site of the former M&S store and office building. Reflected is the Triangle shopping centre (the former Corn Exchange), which was also badly damaged by the bomb in June 1996.

- At the time of the passage of the Town and Country Planning Act, the world's first computer was being developed and tested at Manchester University. There was probably no significant connection made between these two things until the late 1960s and early 1970s with its relatively short-lived fashion for large-scale computer modelling, but since then (and as that fashion has waned) the understanding of how the electronic revolution would both affect cities and reinvent the nature of information flows between the state and its citizens has grown rapidly (Castells, 1989; Graham & Marvin, 1996).
- The concept of globalisation as the driving force of much economic change was not understood in the late 1940s. It can be argued that British planning initiatives were slow to understand not just this major economic trend but also the differences between those economic forces that they could not control and those forces where significant influence was possible (see, for example, Savitch's reflections on planning in London, which are very pointedly entitled 'Ambivalent Development, Ambivalent Result'; Savitch, 1988, Chapter 6). Economic change was regarded as an endogenous variable, giving rise to mechanistic advice about how industrial land

needs should be calculated which translated an assessment of future jobs needed into land allocations through average job densities per acre (see, for example, Keeble, 1964, pp. 187–190). Many locations still have (empty) industrial land originally allocated in their development plans on this basis. The contemporary view that economic change should be seen primarily as an exogenous variable (Hayter, 1997), and the consequential view that attracting inward investment and growing indigenous companies both involve a great deal more than land allocation, has taken a long time to establish itself in the British planning lexicon.

- Social change, too, is a profoundly different phenomenon today from the perceptions of the 1940s. The prevailing view then was that housing need should be thought of primarily in terms of the conventional family ('dad, mum and 2.4 kids'), whereas it is recognised today that the primary drivers of housing demand are the growing numbers of one- or two-person households (see, for example, DoE, 1996). Structural changes of this nature have been accompanied by other major social changes as well, such as the drive for gender equality, the growing recognition of the needs not only of disabled people but also of elderly people in a society where longevity is becoming more common, and the development of Britain as a multi-racial society (Thomas, 2000).
- The 1947 Act did not really conceive of the notion of public participation in the planning process, and it was not until the Skeffington Committee report (Skeffington, 1969) that the need for this was formally acknowledged. So the slum clearance process in Britain's major cities, which typically ran from around the mid-1950s to around the mid-1970s, usually proceeded with no effective public participation in the process. For example, Manchester demolished approximately 100 000 houses in the name of slum clearance and built around 80 000 replacements (approximately one quarter of which were built on overspill estates beyond the city's boundaries) during this period, not only without much meaningful public involvement but also until the end of this period without much visible public protest (Manchester City Council, 1995, pp. 23–25). This is the equivalent of demolishing and then building a medium-sized town over a 20-year period; and it is worth reflecting on the fact that a local authority machine was able to do this relatively unfettered less than 30 years ago. Today, this would be inconceivable.
- The long-standing problem of regional imbalance has created a situation where the development pressures faced in English regions can be of diametrically opposite kinds (ODPM, 2003a). So, for example, some parts of England face major problems of housing

abandonment whilst others face a need to identify large amounts of land for new house-building. Whilst this is true when comparing whole regions, it should also be noted that there are often important differences in these terms between localities within the same region, often not separated by great distances; so, for example, housing hot-spots in north-east Cheshire sit very close to some of Greater Manchester's most economically and socially deprived localities. Alongside this issue of regional imbalance in terms of development pressures has been the notion of devolution to Scotland, Wales and Northern Ireland pursued by the Labour Government since 1997, which seems likely over time to emphasise growing differences as compared with the previous systems that were more uniform (Tewdwr-Jones, 2002). As referenda to test public opinion about the creation of regional assemblies in the regions of northern England are rolled out from 2004 onwards[3] (ODPM, 2004), it may be that this pattern of differences will be repeated as between the English regions as well. The development of a European dimension to planning in the wake of Britain's accession to membership of the European Union in the early 1970s can also be seen in this light (Williams, 1996), especially when it is remembered that the mantra for much European thinking about planning is the notion of a 'Europe of Regions'.

These six major contextual changes that have occurred during the second half of the 20th century (and this list is by no means complete) should be sufficient to suggest why a system born in the immediate aftermath of World War Two would be likely to need review by the start of the 21st century. That being the case, there are four key planning concepts that are central elements in the raft of ideas that underpin this system (two of which were built into the 1947 Act, and two of which have in effect been added on to the system it created subsequently) that should also be looked at carefully in the light of these major contextual changes. These are the concepts of certainty, speed, public involvement and sustainable development (Kitchen, 2002).

The idea that the planning system ought to deliver certainty to its users (and particularly to applicants for planning permission) can be traced back to the 1947 Act. 'Certainty' in this context is defined as the ability to forecast the probable outcome of a planning application as a result of consulting the development plan. The contemporary question that arises is whether this can still be seen as being the same concept

[3] The date for the first referendum having been fixed for 4 November 2004.

in terms of its deliverability at the beginning of the 21st century as it was when that Act was being formulated. Given the intervening trends of the kind described above, the present author's shorthand response to this question is 'no'. These and other trends have contributed to the world being a much less certain place than it was, and the planning system has to be able to adapt to this. This the British system has struggled to do, especially if we define certainty at the site-specific scale. This has been reflected in the way in which development plans have tended to move away from the site-specificity of the 1947 Act towards a greater emphasis on trying to get strategic policies in place that have a chance of surviving as useful guides for development. This is not a recent diagnosis either. The Planning Advisory Group, in reviewing the experience of 1947 Act development plans, said the following (Planning Advisory Group, 1965, para 1.23):

> 'The initial development plans produced in the early years following the 1947 Planning Act meant that for the first time the whole country was covered by relatively detailed plans setting out the planning policies for each area. But it has proved extremely difficult to keep these plans not only up to date but forward looking and responsive to the demands of change. The result has been that they tended to become out of date.'

This present author's reflections 30 years later on the performance of the development plans that arose from the review carried out by the Planning Advisory Group were broadly similar. If a minimum test of the utility of plans was that they should have a span of useful life at least as long as the time period taken to prepare them, the suggestion was that most plans would not pass this test (Kitchen, in Tewdwr-Jones, 1996, p. 133). As is argued later, the present Government's criticisms of development plans are also of the same ilk. So the thrust of the argument here is that, whilst it is very understandable that developers and their agents would want development plans to provide them with site-specific certainty, the system has often struggled to achieve this objective. This may be a failing of the system, but it may also be a failing of (or at least a lack of clarity about the meaning of) the concept. Perhaps what experience shows is that the planning system is more likely to be successful at creating strategic certainty (that is to say, certainty about the broad direction of change) than it is at achieving site-specific certainty (Kitchen, 2002).

The problem of speed may be similar, in the sense that its significance as a concept needs to be re-thought. This is not an argument for inefficiency, and there have been extensive attempts to try to make sure that the procedural aspects of work determining planning applications and preparing development plans are as expeditious as possible. The key question is around the relative weight that ought to be attached to considerations of speed. A simple illustration will hopefully demonstrate some of the key issues. If a building can be assumed to have a life of 60 years, an 8-week period for determining the planning application that allows that building to be constructed is the equivalent of approximately 0.26% of that life. If that determination period were to be extended to, say, 10 weeks, the equivalence would be approximately 0.32%. How significant is the difference between these two scenarios of 0.06% when seen in the context of the whole life of the building? Most people's responses would probably be 'not very', although in recording this view it is important to stress that the developer's perspective is not likely to be this at all but to be one concerned with issues such as risk and cash-flow at a time when the project is costing money and is some distance away from generating income. How does the perspective then change if the difference between these two time-frames is the difference between a poor decision taken in a rush and a much better-quality decision taken after careful consideration of all the issues? Again, most people's response would probably be that getting it right is more important than getting it quick, especially since a 60-year life represents a long period of time to have to live with a mistake. The point here is simply this; speed is a component of the service the planning system ought to be giving its customers, but this needs to be speed that is commensurate with good-quality decision-making. The speed at which the system operates, which has (properly) been the subject of much criticism, (see, for example, DTLR, 2001, para 2.4) must not be elevated to the point where it becomes a false god.

Many of the points about public involvement in the British planning system over the past 30 years or so have already been made above. The additional point to note in this present context is that public involvement takes time; and indeed, the greater the scale of public involvement that is attempted, the more time and effort that is likely to be needed. This links back quite directly to the above discussion about speed of decision-making, since in many ways it can be argued that developments in public involvement have played a part in slowing the planning system down (see, for example, Arup *et al.*, 2002, p. 50). This statement is not made in order to suggest that public involvement is an undesirable thing, but even if some operators in the planning system were to draw this conclusion and therefore to restrict its scale, it is unlikely that this would be tolerated at least for any period of time by the public whose involvement was being restricted. One of the reasons why this would not be tolerated today is that the general public is much less acquiescent in its attitude towards the work of government, at all levels, than it once was. The willingness to accept that 'the state knows best' seems to have been replaced by considerable suspicion about the quality of much public decision-making. So this is another element where the lesson of 50 years of experience is that we need to re-think and to re-express basic concepts (in this case, that of speed) in the light of realistic contemporary expectations.

Concerns that the planning system has a central role to play in the achievement of sustainable development are a phenomenon of the last 10 years or so (DETR, 1997, paras 4–7). In effect, they have been grafted on to the planning system that has its origins in the 1947 Act on the assumption that this can be achieved without the need to change that system. The difficulty with this may not be around its inherent desirability but around two elements in particular. The first is that it is much easier to talk in rhetorical terms about the desirability of, for example, using planning systems to control the location of activities so as to reduce the need to travel (DETR, 2001, para 3) than it is to specify how in practice that desirable end will be achieved in a given set of circumstances. The second is that it is not at all clear that key sectors of the population (such as the car-owning public, or many businesses) are actually signed up to this concept if it means changing their own behaviour. This leaves the planning system operating in something of a vacuum.

On top of all of this, there is clear evidence (Arup *et al.*, 2002) that the resource base available to local planning authorities to enable them to deliver their services in the face of changing circumstances and expectations has simply not kept pace with needs. This is not merely a phenomenon of the recent past either; it can be traced back at least as far as the cuts imposed upon local government by the Thatcher Government of the 1980s (see Kitchen, 1997, pp. 13–16). The ongoing effect of this long-term process of under-funding would have been significant in itself but, as has been argued above, the planning job has also become more complex over the years. As a result, these elements have created real problems for the local government planning service in terms of its performance, its reputation with its customers, and (by circular causation) the morale of many of the people who are trying to operate the system.

Taken together, these arguments suggest that a planning system that is not reformed to take account of changes on the scale described above, would be a planning system programmed to fail. The important question may not be so much, 'Why is the Government seeking to reform the British planning system now?' but rather 'Why has this reform process been such a long time in coming?'

6.4 The Government's case for modernising the British planning system

The Government's intention to 'modernise' planning became clear from a Ministerial statement of January 1998, a few months after New Labour had won the 1997 general election. Interestingly, the stated reasons for doing this are described as 'missing dimensions', to be put in place whilst persisting with the basic principles of a 50-year-old system which have 'served us well' (DETR, 1998). They bear only a relatively limited relationship to the analysis presented above of major contextual change and were:

- the European context for planning in this country
- clearer statements of national policy for the small number of projects where decentralisation of decision-making is not possible
- effective arrangements for regional planning so that more issues can be resolved at this level
- a continuous search for improvements in local efficiency
- a willingness to consider economic instruments and other modern policy tools to help meet the objectives of positive planning.

The analysis of the failings of the extant planning system presented in the Planning Green Paper of 2001 (DTLR, 2001, Chapter 2) was more directly critical. In essence, it said the following about the failings of the system:

- It is too complex, too remote from too many people, often hard to understand, and also difficult for many people to access.
- It is often seen as a set of rules designed to prevent development rather than a framework for achieving the development that is socially desirable.
- It is too slow, particularly in terms of the time periods that development plans take to wend their way through all the steps in the statutory process, and the outcomes of the system are too often unpredictable.
- Although the system is seen as having quite a consultative style, it doesn't succeed in engaging communities in the sense that they feel that they are shaping what will happen in their localities.

- The system is not seen as having a customer focus, particularly by the business community, and it is not seen as both offering and delivering clear and high standards of service to all of its customers.

The Planning Green Paper then proceeded to assert that these problems could be tackled by changing the system, advancing a series of proposals as a basis for public consultation. There is scope to argue that system change was likely to be limited in its ability to tackle this agenda by itself, without a more thorough consideration of some of the contextual changes that had affected the system throughout its 50-year life and some re-thinking as a result of what would be realistic expectations of it in the contemporary world (Kitchen, 2002); but very little of this is to be found in the Planning Green Paper. Interestingly, the Government's response to the public consultation generated by the Green Paper (ODPM, 2002b), whilst reiterating its commitment to the major changes proposed, begins to acknowledge that changing the system by itself will be insufficient if its objectives for the system are to be achieved. As a consequence, alongside the proposed systemic changes it promotes the need for 'culture change' (ODPM, 2002b, paras 67–86). The next two sections, therefore, look at the proposed system changes and then at this associated concept of culture change.

6.5 The Government's system – change proposals

The Government's proposed system changes can be looked at broadly in terms of three spatial scales: the national, the regional and the local. The five key objectives for the planning system that underpin these changes are seen by the Government as being as follows (ODPM, 2002b, para 3):

- deliver in a sustainable way key Government objectives such as housing, economic development, transport infrastructure and rural regeneration whilst protecting the environment
- create and sustain mixed and inclusive communities
- be transparent so that the right decisions are taken more quickly, with a set of rules that everyone can understand
- enable local communities to be involved much more positively than before
- deliver a higher quality and better respected public service.

The essential features of the changes proposed at national level are (ODPM, 2002b, paras 16–24):

- 'Reviewing national planning guidance, and particularly the content of the Planning Policy Guidance note system (PPGs), to separate out what is genuinely national planning policy from what is simply good practice advice.
- Improving the performance of the Planning Inspectorate (the "arms-length" body that deals on the Government's behalf with planning appeals) through tightening targets, ensuring that appropriate resources are available to tackle the job to be done, and encouraging users of the system to co-operate with the desire to achieve performance improvements through means such as choosing cost-effective routes through the system.
- Improving Government's own performance in dealing with the cases on which it has to make decisions.'

At the regional scale, the main proposed changes (ODPM, 2002b, paras 27–31) are:

- Existing systems of Regional Planning Guidance (RPGs) will be replaced by a new duty on the part of regional planning bodies to prepare Regional Spatial Strategies (RSSs).
- RSSs will be about establishing specific regional and sub-regional policies as well as about translating national planning policies into their particular regional contexts.
- RSSs will be given statutory status (they will in effect become part of the development plan), and will have a particular role to play in providing a long-term planning framework for the work of Regional Development Agencies.
- Local Development Frameworks (see below) and Local Transport Plans will need to be consistent with RSSs.
- The potential problems of 'democratic deficit', created by giving unelected regional bodies more statutory responsibilities in relation to the development plan system through requiring them to prepare RSSs, will be resolved by offering the opportunity for a referendum on the creation of an elected regional assembly in each region where it is clear to the Government that there is significant interest in this. A majority 'yes' vote in such a referendum will trigger a further round of local government reorganisation in that region so that local government becomes a single-tier operation throughout the region (ODPM, 2002a).

At the local scale, the main Government change proposals (ODPM, 2002b, paras 32–40, 54 and 55, and 59–63) are as follows:

- Major changes to the development plan system to create Local Development Frameworks (LDFs). These will contain a portfolio of documents appropriate to the local situation, including a strategic statement which is expected to be kept up to date on a regular basis.
- Changes to the development control process which are essentially about improving the efficiency of the process.
- Local government is expected to become more business-friendly and more customer-orientated.
- Local planning authorities will be expected to promote more effective public engagement with planning processes, using as a policy tool the preparation of Statements of Community Involvement (SCIs).
- Most of the developments both in improving public involvement in planning processes and in making information about planning more widely available are expected to take place at the local level.
- All planning services are expected to be on-line by 2005.

As well as introducing a raft of new sets of initials that providers and users of planning services will need to get used to, this puts in place quite a challenging agenda for planners. It is saying that it is up to the people who operate the system to do their best to make it work, against a background where for years the pre-existing system has not been working effectively. To the extent that this is indeed about changing the system, the professional planning community has begun the process of thinking through how to respond to this challenge (see, for example, Planning Officers Society, 2002, 2003). But it is clear that changing the system will not tackle problems such as an inadequate resource base for local government planning services, poor morale amongst many of those who provide such services, and a low level of regard for the planning system amongst many politicians, developers and members of the general public.

The recognition that a set of issues of this nature needs to be tackled alongside measures to change the planning system if the Government's five key objectives for that system are to be achieved has become a much stronger feature of the more recent documents from Government about the planning modernisation agenda, around the broad concept of 'the culture of planning.'

6.6 Changing the culture of planning

There is room to argue that the concept of 'changing the culture of planning' has become a grab-bag for everything that needs to be changed that isn't amenable to system change via an alteration to the law, and that as a result it suffers from a lack of clarity as a concept. There is also scope to argue that it is difficult to define what 'the culture of planning' actually is as the first step in changing it. One way into this is to look at dictionary definitions of 'culture', and then to try to apply them to planning. So, for example, the 1998 version of the New Oxford English Dictionary offers two relevant definitions of 'culture' in this context:

> 'the customs, arts, social institutions, and achievements of a particular nation, people or other social group'

and

> 'the attitudes and behaviour characteristics of a particular social group'.

It would probably be difficult to define the culture of planning precisely in either of these senses, given the very broad range of people who could properly be regarded as stakeholders in the contemporary planning system. For the present author, this very real difficulty is not as significant as the fact that Government has (perhaps belatedly) grasped the point that its objectives will not be achieved by changing the formal system alone, and has used the concept of 'changing the culture of planning' as a means of labelling this recognition. So, if a degree of imprecision of expression has to be accepted for now, as the price to be paid for the recognition that there is a broader agenda to be tackled in order to increase the likelihood that the British planning system can be programmed to succeed, then in this author's view that is a price worth paying.

The Government's conception of the key issues to be tackled under this heading of 'changing the culture of planning' has five primary elements to it (ODPM, 2002b, paras 67–86):

- Improving the clarity of visions of the role and purpose of planning.
- Setting clear and challenging targets for the key aspects of service delivery via the 'best value' regime.
- Ensuring that the appropriate tools are available for the job to be done.
- Developing and extending planning skills and knowledge, not just amongst planners but also amongst other key players in the system (for example, elected members who participate in planning decision-making, developers, and the general public at large in terms of improving their ability to engage meaningfully with the system in operation).

- Ensuring that appropriate resources are available to local authority planning systems so that they are able to deliver to acceptable standards.

This agenda was clarified in a Ministerial statement in March 2003 (McNulty, 2003), which listed four key tasks under the 'culture change' heading to which Government was committed:

- Set out a vision for planning which can inspire and engage (all stakeholders in the system, that is, and not just the planners).
- Deal with the issues of skills, to enhance capability and morale (starting with the capacities needed on the part of the people who will operate the system).
- Raise the profile of planning in local authority policy making and governance.
- Promote and encourage community involvement in planning.

These are challenging tasks, the importance of which can be illustrated by posing a very simple question: what chance do we believe that the system changes proposed in the Planning and Compulsory Purchase Bill will have of achieving the Government's stated objectives for the planning system if these four tasks are not all successfully addressed? This author's response to the question can be summarised in two words: very little. So this culture change agenda is vastly important. Changing the planning system may well be a necessary condition of achieving the Government's objectives for planning, but it certainly isn't a sufficient condition.

6.7 The challenge of change

A useful starting point for thinking about the challenge of change is to remind ourselves that planning systems are social constructs. The British planning system takes the form that it does because a social need for it was identified, because Governments via Parliamentary processes have put in place legislative frameworks designed to turn that social need into a reality, and because many individuals in many walks of life have played their parts in operationalising this up and down the country (Ashworth, 1954). That reminder should tell us that the changes to the system are likely to be as good (or as limited) as the people who will in practice have to make them work. If anything, this should reinforce still further the significance of the culture change agenda noted above, and remind us of the importance of addressing this effectively if the initiative as a whole is to succeed.

A second point, which has already been made but which bears repetition, is that attempts at changing the British planning system have been undertaken before. The history of these previous attempts is that, although they appeared to be succeeding for a while and were initially quite widely welcomed, over time they tended to struggle with essentially the same recurring problems. This may well be because the contextual changes have been sweeping, indeed pervasive, in their nature, and that attempts to reform the system have struggled with them because they have not fully re-thought the job the planning system was expected to perform. A system fit for purpose in 1947, when looked at like this, is one inherently unlikely to be fit for purpose in the vastly different world of 2003. This observation makes the case for change but also suggests that history is at risk of repeating itself without a contemporary view of what British society expects from the planning system, and unless we ensure that the system is given the best possible chance to succeed. The willingness to address the 'culture of planning' agenda in parallel with changing the system does suggest that the 2002–2004 round of reforms may have a better chance of succeeding than (with the benefit of hindsight) some previous attempts. It will be important for the culture change elements of this reform process to be pursued vigorously for what might need to be a considerable period of time, and not to be quietly dropped once the new legislation is in place, or ministerial champions move to other positions.

Some key questions might be as follows:

- To what extent might there be a risk that so much change all at one go will de-stabilise the existing system and actually make the process of achieving performance improvements more difficult? After all, a typical change process involves at least three phases: sorting out what needs to be changed, implementing that programme of change, and then getting the changed system up to full operational speed. It can be very difficult to operate 'business as usual' in terms of the ongoing process of service delivery when effort has also to be put into all three of these phases. In other words, the process of change can itself further worsen the very problems to which it is trying to respond by undermining performance during this interim period, and thus can lead to a lowering of the baseline position from which the newly-reformed system is starting. This difficulty can get magnified by the fact that change processes tend to concentrate on what is wrong with a system (since that is the

reason for changing it) and to under-emphasise the strengths of that system, whereas the level of performance in the transitional period will depend on those strengths. This is not an argument against change *per se*, but it is a reminder about the need to handle the process of change carefully. Is this getting the attention it deserves?
- The formal system can be changed over a limited period of time via parliamentary processes, but changing the culture of the system is likely to be both a much more protracted and a much less clearly focused process. Are we sure that the momentum needed for this latter task is there? Is there sufficient understanding of the nature of this task, and sufficient buy-in to its necessity on the part of key stakeholders, to enable it to be persisted with when the going gets tough?
- Do we have a 'modern' understanding of how we want the balance between issues such as speed, certainty of outcome for stakeholders and effective community involvement to be struck by this 'modernised' system? Do we have a clear idea of where the drive for sustainable development sits in this balancing process? Since this balance may very well not be the same everywhere, do we have a clear understanding of the roles that the three different spatial scales of decision-making (national, regional and local) should play in striking it? The Government's rhetorical position appears to be that, for example, more effective community engagement can be achieved at the same time as improving speed of decision-making, and it is easy to see why this would be a very desirable outcome if it could be achieved. But are these always going to be mutually-compatible objectives? And if they aren't, what do we see as being the relative priority that should be attached to each, and how should this be determined?
- How will we know how well we are doing in this process of change? One message of the culture–change agenda is clearly that the system needs to be programmed to succeed in order to deal with issues like resources, morale and reputation, but we have to know what this will all actually mean if we are to measure progress. Do we have a clear understanding to be able to know not only whether the reform process is on the right track, but also how to adapt the process of change if it becomes clear that this is necessary? Are we even clear enough about what our current starting points are to be able to measure change from them reliably?

Other issues could be raised under the banner of the challenge of change, but for this author those above are amongst the most important. They suggest that it is what local planning authorities in particular, and planners in general, do with this change opportunity that will matter, rather more than whether the specific legislative changes are themselves 'right' or 'wrong'. It is worth remembering in this context that in terms of reputation, both amongst planning professionals and their immediate customers, some local planning authorities are properly regarded as having been very successful and others (equally properly) are seen in a less favourable light; in these terms, British local authority planning practice is not a level playing field. So perhaps the most important challenge is the need to recognise, to understand and to disseminate why this is so, to identify from this process what is reliably regarded as good practice experience, and to concentrate on raising performance levels particularly amongst the lesser performers in this light. If we could raise the overall average by these means, that would be an achievement indeed.

Checklist

- Major changes are taking place in the planning system; consider whether or not these are likely to affect the proposed redevelopment.
- Increased public involvement in planning applications is proposed – it may therefore be appropriate to consider involving community and/or business groups in the project before submitting the planning application.
- All planning services are due to be on-line by 2005.

Chapter 7
Geographical Information Systems

James Cadoux-Hudson & Donna Lyndsay

7.1 Introduction

Most people are comfortable with using computers to access electronic information in the work place or at home. Such systems store a variety of information or data about something or someone and in the context of previously developed land this does not exist as a single dataset. Much of the data is map based, which can be analysed in a variety of ways within a Geographical Information System (GIS). The power of GIS lies not only in its ability to analyse and display data referenced to the map but also in its function as a database that can bring spatial data together from a wide range of sources (Wyatt & Ralphs, 2003, p. 9). Numerous organisations and companies use GIS in this manner, including local authorities, commercial search organisations and government bodies.

GIS differs from conventional mapping in several ways. Maps were always designed by cartographers for specific uses, such as recording land use or for risk assessment as in the case of insurance maps. The data available were interpreted by cartographers, who would find the best method to generalise the information and to represent it for the intended use. As each map was designed for a particular purpose it was very time consuming to analyse or combine information from the various maps. On the other hand, GIS is a software tool that combines a database management with mapping, through use of the following subsystems (Wyatt & Ralphs, 2003):

(1) A data input system that collects and pre-processes data from various sources. Such systems usually include the facility to restructure data and perform editing operations on digital maps, imagery and databases.

(2) A data storage and retrieval system that organises the data in an efficient manner and allows for retrieval, updating and editing.
(3) A data manipulation and analysis subsystem that performs query and analysis tasks on the data, combining and comparing datasets, statistical analysis or modelling functions.
(4) A reporting/presentation subsystem that displays all or part of the database in tabular, graphical or map form and allows for the production of customised hard copies or output files.

Information contained in a GIS comprises 'spatial data' – geographic data and 'attribute data' – the values that apply to the selected spatial data (e.g. the unemployment rate for an area, or the number of patients in a hospital at a point). The 'attribute data' are stored in 'layers', which allows the spatial elements of the data to be explicitly manipulated particularly for display purposes (especially mapping) and analysis. A GIS holds a central repository of all of the required layers of information, which can then be displayed or analysed by the user to create their own customised maps or to obtain information. A typical GIS for use in connection with previously developed land might comprise one or more 'base maps', overlaid with layers of data, which might include site areas, building areas, number of floors in buildings, site ownership, planning history, current and past uses, discharge consents, abstraction licences, landfills, etc.

GIS links a geographic component – i.e. a coordinate – with the data (information). The data can also be displayed so that the spatial relationships can be visually identified or the system itself can calculate spatial relationships between data, e.g. the measurement of distances between objects, or the areas of land and buildings.

> 'There are two commonly used data structures in GIS – vector approaches and raster approaches. In the vector approach, objects on the ground are represented using three simple building blocks: points, lines and areas. Line and area features are composed of linked sets of points and objects like houses or parks are modelled by connecting points together with straight lines and grouping the lines into areas. In the raster approach, the area of interest is divided up into tiles or grid cells of equal size and geographical extent. Cells are then coded according to the properties of the area they cover.' (Wyatt & Ralphs, 2003, p. 39)

As with any tool, especially a computerised system, a GIS will only produce good results if the data used are of a quality commensurate with the end result and that the functions used are statistically or mathematically sound. The principles of GIS are more fully described in the Ordnance Survey online summary in their 'GIS Files™' available from the Ordnance Survey website. There are many examples of GIS in action but few relate to previously developed land.

7.2 GIS and previously developed land

To enable an understanding of any previously developed land it is essential for a site assessment to be undertaken (see Syms, 2002, Chapter 5), to determine the history of the site, any potential contaminants, any environmental constraints, natural geology and topology, current land use and so on. Undertaken manually for each individual site such assessment is hard to achieve, as many different sources need to be consulted. With digital data and using GIS as a geographic search tool to bring all the data into one place it is now much easier. Such tools are now quick to operate and are relatively cheap to obtain; however, collation of the data is the most difficult part.

Government initiatives such as NLUD (see Chapter 1) record information on previously developed land but much of the information held is inconsistent and incomplete, as it relies on 'grossing up' to overcome the problem that not all local authorities provide the necessary information. It is critical for an intending developer to gain an understanding of the history of any designated site and this can only be based on real data. Virtually all contamination arises from past land use and the only consistent national record of past land use in the UK is provided by Ordnance Survey large-scale mapping. These OS maps go back to the middle of the 19th century. Detailed analysis of historic maps is a laborious task and requires not only the sourcing of the maps, which are available from a variety of sources, libraries in particular, but also requires the relevant features to be manually identified and logged and information moved from the old projection (Cassini) to the current map Projection (Transverse Mercator).

An alternative to this is to obtain the data from commercial organisations that have undertaken both the scanning and capturing of these maps and associated information. These maps have proved useful as a basis for the analysis of land-use information undertaken by local authorities to fulfil their obligations under Part IIA of the Environmental Protection Act 1990. Similarly, a developer should be aware of all historical and current activity and take this into account in the plans.

7.2.1 *Historical Ordnance Survey Maps*

Before looking at the information held on the Ordnance Survey Maps it is important to understand their limitations. Ordnance Survey was established in 1791 and it gradually mapped Great Britain at a scale of one inch to the mile. It wasn't until the 1850s that more detailed mapping became available and consequently more useful. Maps are only as good as the point in time they represent, often the published date, but it should be understood that the land will have been surveyed two or more years previously. As a result the land-use mapping is not of a uniform date.

In rural areas, where the rate of change is lower than urban areas, fewer maps have been published, leading to a greater potential gap in between the surveys, which can be as much as 50 years. Conversely, urban areas that experience a high degree of change will have been surveyed at shorter intervals, leading to more frequent mapping and consequently a higher degree of confidence that individual land-use changes have been recorded. Regardless of possible shortcomings due to the intervals between surveys, the historical mapping created by Ordnance Survey is the most consistent record of land use. In 1996 Landmark Information Group and the Ordnance Survey identified a need for a national source of historical maps in digital form and subsequently formed a joint venture to hold the maps in an easily accessible form, by creating a digital historical mapping archive of mainland UK based on 1:500 (there were a number of smaller scales) 1:1250, 1:2500, 1:10 000 and 1:10 560 scale maps. The historical mapping archive begins with the County Series maps first surveyed in Lancashire in September 1841. The rest of England, Wales and Scotland were surveyed in subsequent years. Each county was then revised between three and five times prior to 1945.

The number of maps produced at 1:500, 1:2500 and 1:10 560 in the pre-war period, covering four periods of resurvey or 'epochs', are shown in Box 7.1. Town Plans data were also captured from the archive at 1:500, 1:528, 1:1056, 1:2640 and 1:5280 scales. These are highly detailed but limited in coverage.

These early maps were created using a Cassini projection. Each county had a separate Cassini projection, often with different origins (meridians of longitude) from their neighbours; as a result distortions were prone to occur at county boundaries. In 1944–45 the origin was standardised for the whole of Great Britain and mapping was transferred to the National Grid based on the Transverse Mercator projection, which employed a single point of origin at latitude 49° north

Box 7.1 County Series historical maps.

1:500 – 1:528 scale	
Epoch One: first County Series	(1850–1893) 11 569 map tiles
Epoch Two: first revision County Series	(1891–1912) 1789 map tiles
Epoch Three: second revision County Series	(1904–1939) 148 map tiles
1:10 560 scale	
Epoch One: first County Series	(1841–1893) 1028 map tiles
Epoch Two: first revision County Series	(1891–1912) 964 map tiles
Epoch Three: second revision County Series	(1904–1939) 96 map tiles
1:2500 scale	
Epoch One: first County Series	(1843–1893) 40 000 map tiles
Epoch Two: first revision County Series	(1891–1912) 35 000 map tiles
Epoch Three: second revision County Series	(1904–1939) 22 000 map tiles
Epoch Four: third revision County Series	(1919–1943) 7500 map tiles
1:10 560 scale	
Epoch One: first County Series	(1843–1893) 6500 map tiles
Epoch Two: first revision County Series	(1891–1912) 10 500 map tiles
Epoch Three: second revision County Series	(1904–1939) 7500 map tiles
Epoch Four: third revision County Series	(1919–1943) 6000 map tiles
Total	150 594 map tiles

at the south of the British Isles; longitude 2° west of Greenwich, the meridian of longitude running more or less through the centre of Great Britain. The result was a seamless map covering Great Britain (see Harley, 1975, pp. 18–23). One of the advantages of the projection is the ease with which the amount of distortion can be calculated and, in addition, scale variation owing to projection can be lessened (Harley, 1975,

Box 7.2 National Grid editions and revisions.

1:1250 scale	
First edition	1943 to 1993 (56 000 map tiles)
First revision	1944 to 1993 (35 800 map tiles)
Second revision	1946 to 1996 (13 000 map tiles)
Third revision	1951 to 1992 (2800 map tiles)
Fourth revision	1953 to 1992 (500 map tiles)
1:2500 scale	
First edition	1943 to 1995 (172 500 map tiles)
First revision	1949 to 1992 (21 800 map tiles)
Second revision	1954 to 1993 (3900 map tiles)
Third revision	1960 to 1992 (600 map tiles)
Fourth revision	1967 to 1992 (47 map tiles)
1:10 560/1:10 000 scale	
First edition imperial	1948 to 1977 (8800 map tiles)
Revisions	1949 to 1996 (8800 map tiles)
Last edition metric	1958 to 1996 (10 500 map tiles)
Total	335 047 map tiles

p. 19). The postwar National Grid maps consist of 1:1250, 1:2500, 1:10 560 and 1:10 000 scale maps. The dates and approximate number of maps found are indicated in Box 7.2.

Completion of the archive through the scanning of all of the postwar infill mapping known as SUSIs (Supply of Unpublished Survey Information), Superplan Maps, SIM maps (Supply of Information on Microfilm), will add around 500 000 maps to the archive. On completion in 2004, this will result in an archive consisting of approximately one million maps, a substantial achievement of national importance.

When comparing maps it becomes clear that the level of detail changes over time. In the earliest maps even individual trees were mapped. Over time the expense of detailed surveys meant that the practice was supplanted by updates from a mixture of aerial photography and field surveys. This led to a reduction in detail, for instance the identification of groups of trees and key features only, and the generic description of a building as 'factory' or 'works' rather than as a specific type of use-related structure, e.g. joinery works or foundry. The database of historic maps created by the joint venture enabled the systematic detailed analysis of this information.

7.2.2 Applying GIS to historical maps

In 1998 Landmark Information Group completed the creation of a unique database of Historical Land Use Data and Potentially Contaminative Industries, linked to the guidelines set out in the Environmental Protection Act 1990 and the Environment Act 1995. The resultant historical map data or Historical Land Use Data (HLUD) has proved essential for investigations in to the existence of potentially contaminated land by many users. For instance, local authorities throughout England, Scotland and Wales have been able to correctly identify and prioritise the key 'sources' of contamination as defined within section 57 of the Environment Act 1995 in order to establish the presence of 'significant pollution linkages' within their specific area. Environmental consultants have also used the map data for desk surveys to determine areas that need further information and examination.

The GIS database of historical land use contains seven layers of data covering the whole of England, Wales and Scotland; six contaminative use layers corresponding to specific time periods and one land-use layer. Into the six potentially contaminative land-use layers, based on each time period or epoch, potentially contaminative uses were categorised and digitised (a full list can be seen in Box 7.3). The seventh layer, the land-use layer, also contains the following features: Areas of Unknown Fill (water), Areas of Unknown Fill (non-water), Former Marshes and Areas Liable to Flood.

Box 7.3 Historic Land Use classification.

Keywords	Historic Land Use Classification
ABATTOIR	Animal slaughtering and basic processing
AIR	Air shafts
AIRPORT	Air and space transport
ANIMAL	Animal by-products (i.e. animal parts) e.g. soap, candles and bone works
BATT	Batteries, accumulators, primary cells, electric motors, generators and transformers
BREW	Brewing and malting
BRICK	Manufacture of clay bricks and tiles, including associated activities, e.g. brickfields; also solitary kilns (other than limekilns)
C&C	Coal storage/depot
CEMENT	Concrete, cement, lime and plaster products; also includes solitary lime kilns

continued

Box 7.3 (*continued*)

CERAMICS	Tableware and other ceramics
CHEM	Manufacture of cosmetics, manure, fertilisers and pesticides, detergents, oil, organic-based pharmaceuticals, other chemical products incl. glues, gelatines, recording tapes, photographic film
COLLIERY	Coalmining (and the manufacturing of coke and charcoal) – areas include associated surface activities in area and coal mine shafts
D GROUND	Disturbed ground >200 m in one dimension
DEPOT	Transport depot: road haulage, corporation yards
DISTILL	Spirit distilling and compounding
DOCKS	Boat-building, wharf and quays, cargo/transport handling facilities – marine or inland
DYE	Dye and pigments
FLOOD	Areas 'liable to flood' – shown as point features central to flooding area
FOOD	Major food processing, includes large dairies. Exceptionally large-scale corn/flour milling
FOUNDRY	Furnaces and metal processing/casting/forges/ smelting – ferro and aluminium alloys – manganese works. Slag works
FUEL	Sale of automotive fuel
GARAGE	Repair and sale of i) cars and bikes, ii) parts, iii) motorway services
GAS	Oil refining and production of gas from coal, lignite, oil or other carbonaceous material other than waste
GLASS	Flat glass and glass products manufacture
GRAVE	Cemetery, modern burial grounds and graveyards
HEAP	Must be associated with relevant industry – incl. spoil and slag – use symbology and associated features to identify heap boundary
HEAVY ELEC	Manufacturing of distribution, telecomms, medical, navigation, metering and lighting
HEAVY TRANS	Manufacturing and repair incl. i) ships, ii) aerospace, iii) rail engines and rolling stock
HM WORKS	Heavy product manufacture – rolling and drawing of iron, steel and ferro alloys – includes major tube works
HOSPITAL	All hospitals including sanatoriums but not lunatic asylums

continued

Box 7.3 (*continued*)

HOUSE	Manufacturing of electrical and electronic domestic appliances
TRANS	Manufacture of cars, lorries, buses, motorcycles, bicycles
LAB	Various. Technical and environmental testing and analysis
LAUNDRY	Laundries and dry cleaning
LIGHT ELEC	Computers, office machinery, business/industrial electrical goods
MACH	Manufacturing of engines, building and general industrial machinery, incl. nuts and bolts, gas fittings, wire rope/cable and ordnance accessories
MAG	Civilian manufacture and storage of weapons, ammunition, explosives and rockets, including ordnance
MARSH	Feature should only be shown when the following criteria are satisfied: i) It is Epoch 4 or later, ii) Land is vulnerable to or has been built on, in a later Epoch
METAL PROD	Constructional steelwork, metal structures and products and building materials
MINE	Areas of mining and single or groups of shafts other than coal, OR not specified – incl. levels, adits, etc. Also areas associated with Mineral Railways
MINERALS	Abrasives, asbestos, etc. and products
MOD	All military establishments incl. firing ranges (if not specified as civilian)
MRAIL	Mineral railways also known as 'tramways' or inclines – NOT including urban passenger 'tramways'
NEWS	Printing of newspapers
OIL	Major oil and petrol storage and all gasometers which are not in gasworks
OUTFALL	Outfalls incl. warm water, industrial effluent, etc. unless directly attached to other feature, e.g. end of sewer pipe
P	Above-ground pipelines other than sewerage
P PROD	Paper, card, etc. products, e.g. packaging
PAINT	Paints, varnishes, printing inks, mastics, sealants and creosote
PAPER	Pulp, paper and cardboard manufacture

continued

Box 7.3 (*continued*)

PIT	Extraction of alluvial sediments (sand, stone, clay, peat, marl and gravel)
PLASTICS	All plastic goods, including building, packages, tubing etc. and the manufacture of tar, bitumen and asphalt
PLATING	Electro-plating, galvanising and anodising
POWER	Electricity generation and distribution, including large transfer stations
PRINTERS	Printing other than newsprint
QUARRY	Quarrying of all stone (incl. limestone, gypsum, chalk and slate) and ores, includes all opencast mining and slant workings – also slate/slab works, flint works, stone yards
RAILWAY	Railway tracks – up to 4 tracks wide or 30 m
REFUSE	Refuse and waste disposal incl. incinerators and sanitary depot
RLAND	Rail sidings, yards, rail wharf, goods depots, station, etc.
RUBBER	Natural and synthetic rubber products incl. tyres and rubber products
SCRAP	Recycling of metal waste incl. scrapyards and car breakers
SEWERAGE	Sewerage, septic-tanks, effluent – incl. all filter beds
TANNERY	Tannery, leather goods and skinnery
TEXTILES	Natural and man-made textile manufacture and products including hemp rope and linoleum
WIRES	Insulated wire and cable for electrical/tel. purposes
WOOD	Sawmills, planing and impregnation (i.e. treatment of timber), wood products, telegraph works, timber yard, e.g. veneer
WORKS	Factory and works – use not specified

Through the process described above, information on some 400 000 old industrial sites, dating back to the 1850s, has been captured. If the area of these sites is added up, it equates to the size of the county of Devon. In addition around 275 000 sites have been identified which are likely to be filled ground, dating from 1850, including old quarries, gravel pits, even canals have been captured, equating to an area the size of Bedfordshire. Where quarries or mining activities are smaller than

100 metres diameter they are represented by a cross. Linear features such as railway lines are also captured. The historical dataset was subsequently supplemented by an analysis of the 1:1250/2500 postwar maps for all the petrol stations, tanks and utility services, because of their importance in the identification of potentially contaminated land, creating a database of over 390 000 tank and energy facility features alone.

7.3 Using a GIS database to assist in the redevelopment of PDL

As illustrated in the previous section, the type of information contained within a GIS dataset enables the identification of filled and disturbed areas of land, for example filled quarries or water features, potentially contaminative activities or industries such as refuse or slag heaps, tanks and energy facilities and past land use such as gravel pits. Because the geographical information in a GIS is captured digitally, i.e. information is taken from the scanned maps in the form of points, lines or polygons, the information can be used for the purpose of automatically identifying any historical land use within any specified distance of the site.

Figure 7.1 is an example of the type of information captured. The large hatched polygons represent coal and lignite mining in 1889. When this information is overlain on recent mapping, as in Fig. 7.2, it is clear that many of the identified quarries have been built upon, indicating that the quarries have been filled. This in turn poses further questions as to the nature of the fill materials and whether adequate provision has been made by the developer to remediate the site.

If a property has been built on previously developed land, as shown in Fig. 7.2, it is important to understand the history of the site, since any unexpected issues could be very expensive to resolve during construction or post construction. One example of this is an unfortunate property owner whose bungalow was built on an old gas works. When excavations were dug for garage foundations a tar pit was found, resulting in the removal of 600 tonnes of coal tar at a cost of £30 000. If the history of the site had been known prior to purchase it might have been possible to negotiate a price allowing for any issues that might arise from the gas site. The site is clearly identifiable in 1911 as an old gas works. There are many similar examples of sites when a more thorough knowledge of the history of the site would have prevented considerable extra costs and considerable delays.

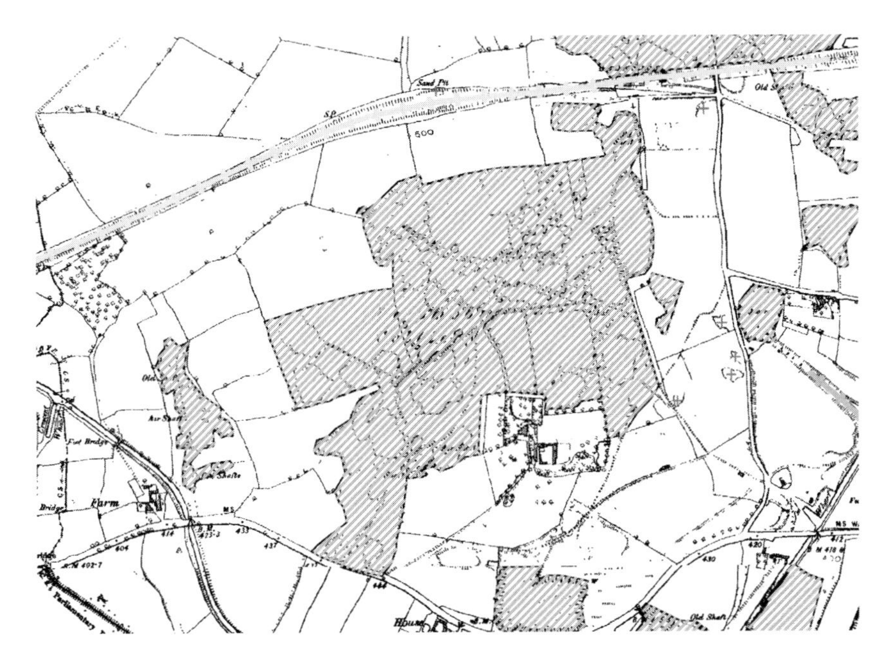

Fig. 7.1 Historical Land Use (shaded area) identified from County Series Mapping. © Landmark Information Group and Ordnance Survey 2003. Reproduced with permission.

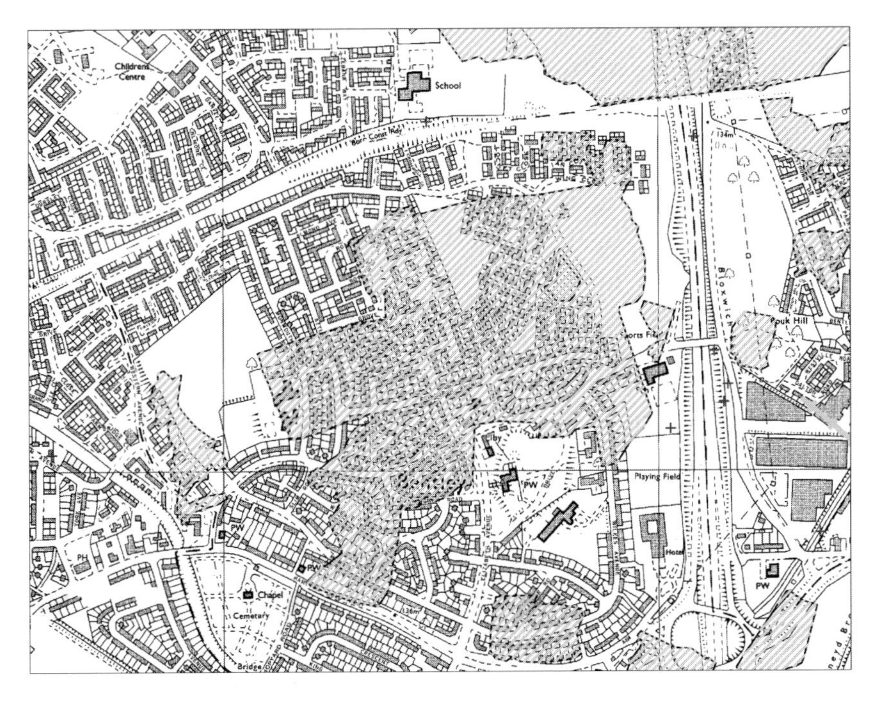

Fig. 7.2 Historical Land Use data (shaded area) overlain on to current 1:10 000 National Grid mapping. © Landmark Information Group and Ordnance Survey 2003. Reproduced with permission.

7.4 Other historical datasets and maps

In addition to historic Ordnance Survey maps, there are other sources of information available that provide varying levels of detail, including Fire Insurance Plans, land-use maps, aerial photographs and landfills.

7.4.1 Fire Insurance Plans

Fire Insurance Plans were developed in approximately 1870 as a result of demands from fire insurance underwriters requiring an understanding of the physical characteristics of building structures that were to be insured. The maps were updated on a regular basis and they enabled insurance companies to consider the risk of fire in areas where there was a heavy concentration of policyholders in order to limit their potential losses. As a result these were only economical in urban areas where the structures were relatively static and, following changes in insurance practice in 1970, creation of the plans was discontinued.

Between 1875 and 1970, the Chas E Goad company dominated the production of Fire Insurance Plans in the UK. The plans were produced in volumes covering 126 urban areas and were leased to national insurance companies. The insurance plans are usually at the scales of 1:480 (1 inch to 40 feet) for the British Isles and 1:600 (1 inch to 50 feet) for foreign countries (see Fig. 7.3). The volumes were updated every 5 years through manually pasting on amendments or through a complete reprint of a sheet; as a result of these update methods the sheets vary in quality. The plans within the volumes depict, through colour and symbols, a wealth of information on land use, internal and external building construction (for example asbestos, concrete), height, street widths, tanks and property numbers and lines.

This is virtually the only source that allows the identification of asbestos materials and internal tanks and is thus significant when considering the redevelopment of previously developed land. The Fire Insurance Plans volumes also contain information on bomb-damaged sites. Other than knowledge of the current GOAD town centre plans, there seems to be very little awareness of these historical and valuable maps: however, the maps are being scanned by Landmark and will be made publicly available in 2004.

7.4.2 Land-use maps

The Dudley Stamp (1931–35) and Alice Coleman (1960s) maps were designed to show specific classes of land use. Because of the nature of

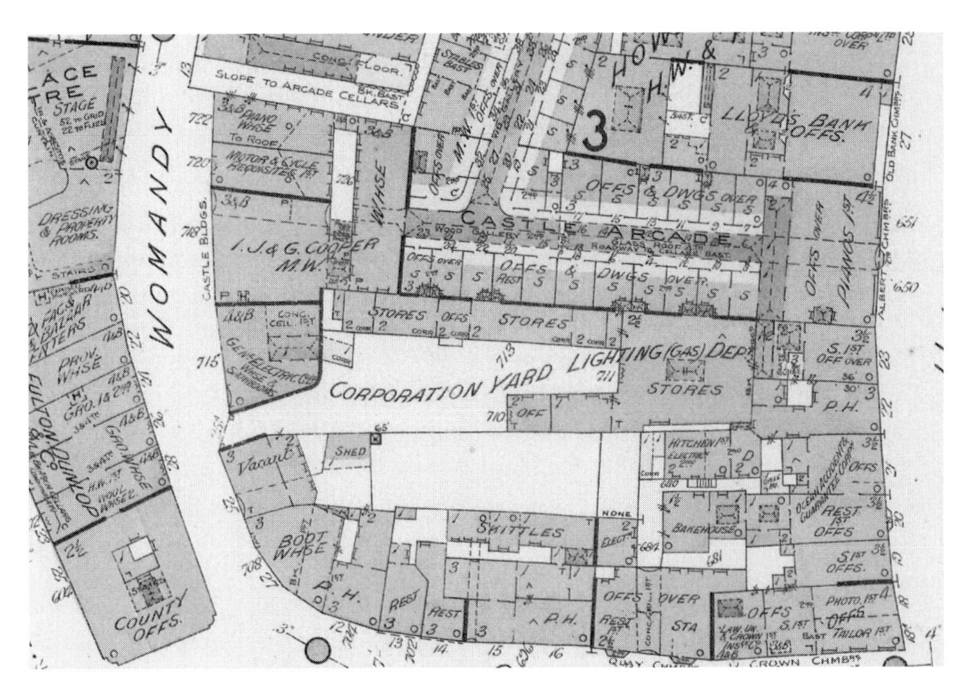

Fig. 7.3 Extract from a Fire Insurance Plan, Cardiff 1909. © Landmark Information Group and Ordnance Survey 2003. Reproduced with permission.

the time difference between surveys across the country neither map provides a consistent image of land use. The scales used are also much smaller (Dudley Stamp: 1:63 360 and Alice Coleman: 1:25 000) than the maps used for the historic land-use analysis previously discussed, making identification of contaminative land use difficult.

7.4.3 *Aerial photographs*

Much contamination also occurred in expanding urban areas in the period after the Second World War. The Ordnance Survey aerial photo maps available from 1945 to 1954 provide snapshots of the land use within the postwar era, although in some editions sensitive military sites were masked by strategically placed clouds or set of fields. These images are available from libraries and the GeoInformation Group. More recent imagery can be obtained from a variety of suppliers including UK Perspectives, GeoInformation Group and Getmapping. The GeoInformation Group and Simmons Aerofilms also hold a range of historical imagery. The barrier to common use of aerial photography is that the images are generally not available in digital form (although there are some exceptions) and they need specialist interpretation. Recent aerial photography has been converted into digital data and is now available nationwide dating back to 1999. Earlier years will not have full coverage.

7.4.4 *Landfills*

In 1973 the British Geological Survey originally conducted a survey of active landfill sites on behalf of the Department of the Environment (now DEFRA). This survey includes over 3000 sites accepting waste prior to the Control of Pollution Act (COPA) 1974, and would therefore not have been subject to any strict regulation or monitoring. Further details which may be available from BGS paper records include outline plans, site descriptions, waste types and tipping histories. Responsibility for landfill used to be with the local authorities but this responsibility was passed over to the Environment Agency when it assumed responsibility under the Waste Management Licensing Regulations 1994. Registered landfill sites can be sourced from public registers held by the Environment Agency (discussed below) and the Scottish Environment Protection Agency, under the Control of Pollution Act (COPA) 1974 and section 36 of the Environmental Protection Act (EPA) 1990. The data held relate to open and closed sites, licensed for the landfill of waste. Sources for landfill today are local authorities, British Geological Survey, the Environment Agency and the Scottish Environment Protection Agency.

7.5 Current sources

Many other organisations have been collecting their data within databases or for the purpose of being held within GIS, triggered by demands for faster access and higher spatial accuracy. The availability of such datasets for the purpose of analysing previously developed land is dependent upon the organisation/body concerned. Commercial organisations source the available data and provide it packaged for use as part of a desktop study. However, it is worth taking each key source in turn as, whilst being extremely time consuming, there are various options for obtaining the information independently.

7.5.1 *PointX*

PointX is a source of current land-use data. PointX has been collecting national points of interest since 2001. Using more automated techniques than those discussed earlier, PointX analysed the text within Land-line®. In doing so, PointX has managed to capture various features that could be potential risks, such as electricity substations and tanks. Further information can be obtained from www.pointx.co.uk.

7.5.2 *Local plans*

Information on constraints set by planners as part of the Statutory Planning process can aid the identification of sites allocated for re-development. The plans are available from the local statutory authority. Landmark has undertaken the capture of such land-use constraints information, creating a dataset representing the information from development plans across the country. Each local authority will have its own classification of policies, which makes such an undertaking relatively onerous. To ease identification of the constraints the information from several thousand classes has been normalised by Landmark to a simple top-level classification, see Box 7.4.

Box 7.4 'Top-level' classification of land uses.

Housing: residential and other housing developments, including redevelopment and conversion.
Transport: transportation including planned corridors for new roads, minor and major road and rail alterations and a diverse range of cycle, pedestrian and parking policies.
Open land: includes green belts; nature conservation areas; Sites of Special Scientific Interest (SSSIs) etc. – constraints normally relate to the preservation of such areas and restrictions on residential or any other development.
Heritage environment: areas of many towns are designated as conservation areas under statutory powers. Additional statutory hurdles will be in place, and policy constraints are also often imposed concerning the types of construction that will be allowed.
Town centre and retailing: includes retail areas, retail servicing areas and town centres.
Industrial and commercial: includes commercial development sites, offices, warehousing, hotels and business and industrial use in general.
Community and social facility: includes health, education, sport, leisure and social areas and allotments.
Waste, pollution, minerals, water and energy: includes waste processing and disposal sites, land use for utility purposes (such as sewerage) and potential development hazard areas.
Rural/settlement: villages and other rural areas, urban areas.

In the assessment of previously developed land the housing category can provide valuable information on areas set aside for redevelopment. Other zones can also indicate where there may be issues in transporting large amounts of rubble or contaminated material away from the site.

7.5.3 *Environment Agency*

The Environment Agency collects information from its offices and stores the data in various databases. The Environment Agency was set up under the 1995 Environment Act. As a non-departmental public body, it is sponsored largely by the Department for Environment, Food and Rural Affairs (DEFRA) and the National Assembly for Wales (NAW). The Environment Agency is legally obliged to make certain information about its licensees and their operations available to the general public (as a statutory remit). This information has been collected into public registers, and it is held in the Environment Agency's area offices. These registers can be easily accessed through contacting the National Centre on 0870 850 6506. Some information is also available on their website at http://www.environment-agency.gov.uk and at the 'What's in your backyard' part of the website. The types of information available on their public registers are shown in Box 7.5:

Box 7.5 Information on Environment Agency Public Registers.

Water

- The Water Quality and Pollution Control Public Register
- The Water Abstraction and Impounding Register
- Maps of Freshwater Limits
- Maps of Main Rivers
- Maps of Sensitive Areas and High Natural Dispersion areas
- The Groundwater Register

Waste

- The Register of Carriers of Controlled Waste
- The Register of Brokers of Controlled Waste
- The Register of Waste Management Licences
- The Register of Exempt Activities
- The Producer Responsibility Register
- The Professional Collectors, Transporters of Waste, Dealers and Brokers Register

continued

Box 7.5 *(continued)*

Industrial pollution

- The Integrated Pollutions Control (IPC) Public Register
- The Register of Radioactive Substances Information
- The Pollution Prevention and Control Register
- The Control of Major Accidents and Hazards Register
- The Contaminated Land Register

Other

- The Maps of Agency Waterworks

7.5.4 *British Geological Survey*

The British Geological Survey (BGS) maps are available in hard copy and digital formats for GIS application. Digital data are available at various scales from 1:625 000 to 1:10 000. Full digital coverage of Great Britain can be provided at 1:50 000 scale. The coverage at 1:10 000 scale is at present limited, but digital data are available for London and many of the major cities. The BGS supplies digital data under a non-exclusive licence agreement that can be viewed at http://www.bgs.ac.uk/products/digitalmaps/home.html. The 1:50 000 digital geology is provided as four separate layers including the solid geology, drift deposits, mass movement and artificial ground of which the artificial ground is the most relevant in relation to contaminated land studies. Maps, publications, reports and data can be ordered directly from the BGS website www.bgs.ac.uk and their online shop http://www.bgs.ac.uk/shop/.

7.6 Conclusion

This chapter has provided a brief outline of the benefits that can be obtained through the use of GIS and the different datasets available. The approaches to pulling this data together can range from the purchase of a standardised report from a commercial provider, such as Landmark Information Group, through to the integration of the GIS and data with appropriate software to assist with the development risk assessment process.

The roles that GIS can play in relating environmental information to the development process are many and are not limited to the identification of former land uses that may impact upon the redevelopment proposals. A great deal of publicity has been given over the last two or three years to the plight of homeowners living in flood risk areas, who find themselves unable to obtain insurance cover for their homes and contents. In an attempt to at least reduce the numbers of properties affected, 'Norwich Union, Britain's biggest insurer has introduced a digital mapping system that enables it to accurately assess the risk to individual homes' (Wright, 2003). This replaces the previous postcode-based system and, for example, Norwich Union has said that it will insure up to 30% of homes in Shrewsbury, where previously cover was not offered. The greater accuracy of the GIS-based approach does, however, have its downside as some properties, still excluded from insurance cover may be even further blighted, although a spokesman for Norwich Union claimed that 'more people will benefit than will lose out'. This accurate mapping of flood risk areas will also be important to developers.

In conclusion, it is important to check as many sources of information, both historic and current, in order to understand the nature and history of a previously developed site. Without undertaking a preliminary site investigation developers are opening themselves to potentially costly problems that may not be picked up on until late on in the process. Intending developers can either source the information themselves or go to a commercial organisation, which can pull together various strands of information about the site. Either way, detailed research is the key to understanding the site. Geographical Information Systems can assist the process both by making information readily available to the development team, that would otherwise be very time-consuming to collate, and through enabling clear-cut decisions to be taken with regard to the development risks.

Checklist

- Does the information about the site include a comprehensive report about historic land uses, pollution incidents, landfills and other environmental data that may affect the proposed development?
- Consider whether or not GIS can assist in the decision making process.

Part B

Industrial Activities and Contamination

Chapter 8
Industrial Activities and their Potential to Cause Contamination

8.1 Introduction

In preparing the information in Part B, consideration has been given to the types of situations which may confront valuers, surveyors and their clients, whether they be occupiers or the intending developers of industrial properties. The situations considered include those such as the interpretation of environmental information, including what to do when presented with desk studies and site investigation reports produced for vendors or third parties. Questions that may have to be addressed include:

> 'The desk study or historic report tells us about the past uses on the site, from the middle of the last century, but how great or relevant is the risk?'

or

> 'This report was prepared for the previous owner; can our client rely upon it?'

and

> 'Should we advise our client to commission an intrusive investigation of the site?'

There are no simple answers to questions of this nature, especially in terms of the risks associated with past or present land uses. Each site is unique in terms of its physical, geological and chemical characteristics, and its history of use. Factors such as site geology, hydrology and hydrogeology need to be considered, together with the period during which the manufacturing plant operated and the nature or age of the plant itself. For example, very different situations may exist in respect to a factory constructed on a sandy sub-soil, on a river bank, which has been in use since the beginning of the 20th century, compared to a modern factory, say less than ten years old, used for the same purpose, in a less sensitive physical environment. Raw materials storage and waste management and disposal practices may also have changed during the operational lifetime of the manufacturing plant.

Hazards should be viewed in the context of the nature of the target considered to be at risk and the point in time at which the risk is to be assessed. Under the contaminated land legislation there has to be an unbroken link between the contaminative substance, the source or *contaminant* and the potential target, or *receptor*, via a *pathway*. Break the link, say by removing the pathway or the receptor, and the land would no longer be contaminated in the legal sense, even though the contaminants remain on-site.

Receptors also need to be considered in terms of the degree of risk arising out of their exposure. For example, different exposure criteria would need to be applied in respect of people who work on the site every day (say in a manufacturing plant), occasional visitors or perhaps ground workers constructing foundations for a new building. In each case the exposure of the individuals concerned will be significantly different. It should be noted that the receptor does not have to be human; harm to property, crops, livestock and protected environments, such as Sites of Special Scientific Interest (SSSIs), are also covered by the contaminated land legislation. It is also good practice to use the risk-based approach to assessing contamination risks in other contexts as well as Part IIA, for example for planning and development control purposes.

In many instances site investigation reports may be provided to the valuer, or developer, to assist in preparing the valuation or arriving at an offer for the site. Consultants acting on behalf of the vendor of a site or an existing industrial facility may have prepared these reports. They may even have been prepared for a previous prospective purchaser and provided to the site owner under an agreement requiring the information from that party's investigations to be made available. Either way, it is most unlikely that the consultants that prepared the reports will owe a duty of care to a prospective purchaser or any other third party. Nevertheless, such reports may provide useful background information, which can be of assistance to the valuer or intending purchaser. The task of the valuer will be especially eased if a SiLC or other suitably qualified person has properly recorded the information from earlier investigations in a Land Condition Record (see Chapter 4).

To be of greatest use to the valuer or developer, reports should, in addition to describing ground conditions, seek to establish the probability of occurrence for the contamination identified and the likely costs of decontamination. They should consider liabilities associated with groundwater, cross-boundary migration and public nuisance. The reports should also assess the likely or actual *source–pathway–receptor* relationships. If these issues have not been addressed, consideration should be given to the commissioning of a new report before the valuation report or purchase is completed. Even if these issues are covered in the reports, there may be gaps in the information, or the purchaser/developer may wish to obtain verification from their own consultants.

Where seemingly comprehensive reports have been commissioned by the site vendor, or some other third party, it will be necessary to ascertain whether or not the consultant is prepared to assign the report to a new owner, provide 'duty of care' or 'privity of contract' to someone other than the commissioning client, by entering into a collateral warranty. The information contained in such reports also may become obsolete, or change over time, and thereby increase or decrease in importance. This is especially important in situations where the reports are more than a few months old and the industrial activities have continued during the intervening period since the reports were prepared, or where the site has been left unsecured, giving rise to the possibilities of fly-tipping of both liquid and solid wastes.

It may be appropriate, if some additional investigation is needed, or the site is to be remediated, to re-appoint the original consultant to carry out the additional investigation or to oversee the remediation work. All appointments should be made by way of a binding contract, on the client's terms, rather than the standard terms or conditions of the consultant. The existence of adequate and suitable Professional Indemnity Insurance, held by the consultant, will need to be considered, as the cover provided by some consultants is very limited in terms of scope and monetary value. It may also be appropriate for a prospective purchaser or developer to consider taking out additional insurance to protect against the possibility of contamination, see Chapter 3.

Given that it would be an impossible task to produce a range of answers to the *where do we go from here?* type of questions, which would cover all eventualities, the methodology used has been to review existing material in the subject area and to produce some general advice which will enable people to answer their own questions. It is therefore most important that prospective purchasers, intending developers and their respective advisers gain a thorough understanding of the activities that have taken place on the site, both in the immediate period leading up to the transaction and over time.

8.2 The potential for contamination

The ways in which land used for industrial purposes may become contaminated are unbounded. It is nevertheless possible to identify three main stages, applicable to most manufacturing processes, which may result in contamination (see Fig. 8.1). These are:

- the delivery, storage and handling of raw materials
- the manufacturing process itself
- the disposal of wastes

In addition, contamination from industrial activities may also be dispersed by air and by water, with the result that contaminants can be spread over a much wider area than the manufacturing site itself.

Industrial property purchasers, including developers and their professional advisers, must bear in mind that the activities with the

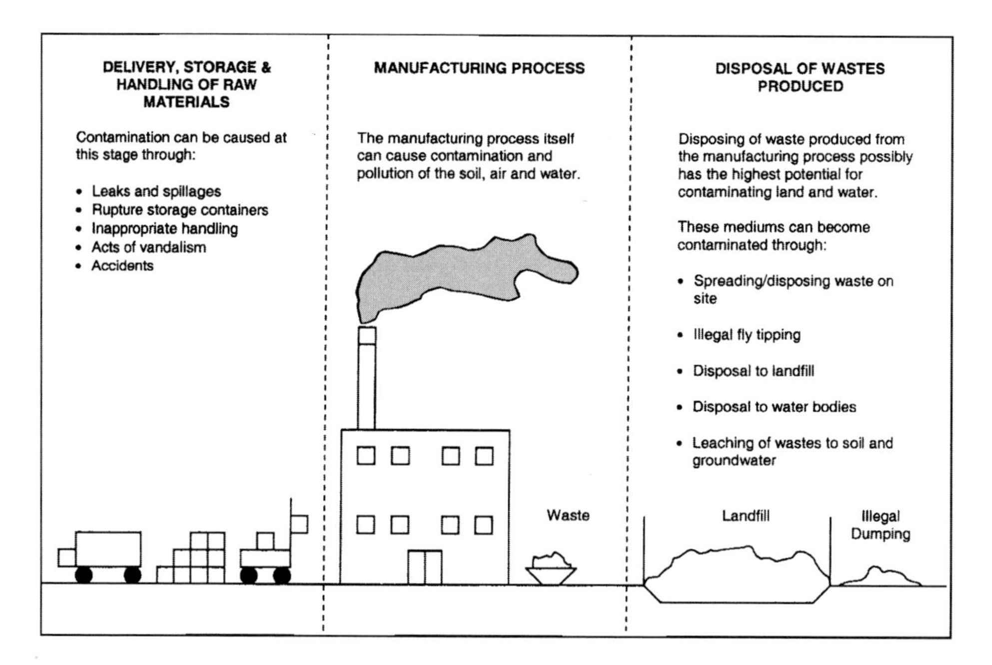

Fig. 8.1 The potential for land contamination through the manufacturing cycle. Reproduced with thanks to Iain McBurnie, final year BSc Environmental Management, Sheffield Hallam University.

potential to cause contamination may not always be located on the land that is to be valued, purchased or redeveloped. The industrial uses may be located on adjoining or nearby land, from which contamination may have migrated. Site redevelopment may also include having to deal with contamination that has been covered or encapsulated during earlier development phases and may not be readily apparent from an assessment of the activities carried out on the site or in its vicinity.

Taking the three main stages of industrial processes, some of the ways in which contamination may occur are described below.

8.2.1 *Delivery, handling and storage of raw materials*

The ways in which raw materials may be delivered to a manufacturing plant, and the containers in which delivery is effected, are diverse and may vary within industries, as well as from industry to industry and over time (see Fig. 8.2). For many industries bulk materials, in liquid or powder form, may be delivered by road or rail tankers (and occasionally by water), or in smaller quantities in drums, flasks, bottles

Fig. 8.2 Chemicals and other liquid raw materials may be stored in tanks and silos outside the buildings. It is important to locate the position of such structures as part of the historical study and to undertake targeted intrusive investigations of these areas.

or bags. Drums, flasks, bags and other smaller containers may be delivered in palletised form, or as individual items. Other materials, for example cellulose for the explosives industry, may be delivered in bales.

Transfer of liquid or powder raw materials may be direct from tanker to storage tanks, above or below ground, or to silos. The construction and materials used for tanks and silos will depend upon the nature of the materials stored, such as their viscosity, corrosiveness or flammability. Pallets are likely to be unloaded and moved within the manufacturing premises by fork lift truck – gas, diesel or electric powered – or by hand truck. Individual drums and bags may be moved by sack trolley or hand truck.

Pre-production storage of palletised materials may be in purpose-designed pallet racking or floor stacking in warehouses. Drums and bags may be stored in racks, storage bins or on the floor. Some raw materials may be stored in the open, in fenced compounds, open sided sheds or other discrete storage areas.

Oils and hydrocarbon based fuels, used for road vehicles, production processes and space heating may be found on sites used for many different purposes. These may be stored in tanks, above or below ground, or in drums, all of which may leak or have resulted in contamination through accidental spills. Solvents are used in many industries, from dry cleaners to tanneries and engineering works. These can migrate through the ground and may contaminate ground and surface waters.

The handling of raw materials within the production facility, from storage areas to the process plant, may be by a variety of methods including pipeline, conveyor, forklift truck and hand truck. Throughout the operations of delivery, storage and handling of raw materials within the production facility many opportunities exist for contamination to occur. These may include leaking storage facilities, bunds, drains and pipelines, accidentally ruptured drums or sacks, other accidents, inappropriate handling or storage of raw materials and vandalism.

8.2.2 *Manufacturing processes*

With the diversity of manufacturing industries covered by the land-use categories described later in Part B, it is not possible to include any information as to the potential for land contamination to be caused during manufacturing processes. Therefore this section is confined to general comments regarding contamination.

Manufacturing processes have the potential to cause pollution of the soil, air and water environments, either through normal process operations or as the result of accidents, including explosion. In order to determine whether or not soil contamination is likely to have occurred it is necessary to consider each property on a site and process specific basis, which will include ascertaining details of site operating procedures, obtaining information as to historic activities and records confirming compliance with conditions in operating licences, permits or consents, as well as industry specific regulations or trade practices. Information relating to prosecutions and other enforcement procedures by environmental regulators will assist in determining the extent of any problems that might exist.

The assessment will need to identify which parts of the site were used for specific activities in the manufacturing process and how these change over time, for example with the introduction of new machines. It should be borne in mind that, even where machines have been mounted on substantial concrete bases, lubricating oils, cleaning solutions and other solvents, including volatiles used in the manufacturing process, can seep around the edges of bases, through expansion joints and into the underlying soil. Therefore consideration will need to be given to the possibility of pathways having been created through otherwise impermeable surfaces.

8.2.3 *Disposal of wastes*

The wastes generated by industrial processes, and their method of disposal, will vary from industry to industry. It is also probable that waste disposal practices will have varied over time, with changes in disposal practices most likely to have occurred since the Control of Pollution Act 1974 came into force.

Liquid wastes may be disposed of direct to the foul sewer, either with or without prior treatment on site. Solid wastes may be disposed of to licensed landfill, to a recycling plant or waste processor. In all such cases appropriate licences, for example under the Waste Management Licensing Regulations 1994, and 'duty of care' certificates should contain information regarding the nature and quantities of the waste materials, and the means of disposal. Both liquid and solid wastes may be recycled within the production facility.

In the past, some industries disposed of solid wastes within the boundaries of the production facility, using wastes to fill voids in the ground, as blinding beneath concrete slabs, or to produce 'hard-standing' for vehicles. The wastes involved included surplus or semi-processed raw materials, reject products, packaging materials and fuel wastes such as ashes, clinker and slags. This type of waste disposal may also have extended to nearby properties, sold or given to neighbouring owners for use as construction or fill materials. The industries involved included asbestos manufacture and products, town gas production, foundries, metal smelting and metal finishing works.

8.2.4 *Transmissions by air and water*

Emissions to air, although now more rigorously controlled than in the past, may have resulted in soil contamination occurring some considerable distance from the source of the contamination and the contamination may persist for many years. For example, several years ago the author was asked to provide advice in respect of excessively high sulphate levels found in soil samples taken from a local authority housing estate located in a large industrial city. Consulting engineers had been asked to advise the local authority as to the feasibility of refurbishing the estate and they had taken the soil samples as part of their survey. Approximately two miles to the south-west of the estate had been a town gasworks, which had ceased operation more than 20 years earlier. The conclusion reached was that the burning of high sulphur coal in the gasworks, carried by the prevailing south-westerly winds, had been the cause of the contamination as the heavier particulate matter cooled and was deposited on the ground.

Authorised discharges are an obvious means by which low levels of contamination may enter surface waters; however, such discharges were not always as well controlled as they are today. As a result of historic discharges from industrial activities, contaminants may be carried considerable distances and may still be found in sludges dredged from the beds of rivers and canals in industrial areas (see Fig. 8.3). In some cases liquid wastes were deposited directly into the ground or into waste heaps; these may have percolated through the soil into both ground and surface waters. Leaking drains, both foul sewerage and surface water drains, are another common source of contamination. It is also possible for liquid wastes to have escaped from production and storage areas, via pathways formed by service ducts or through spaces between structural steelwork and concrete floorslabs, thereby allowing contamination to spread beneath apparently impermeable floors and to the underside of stanchion bases or foundations. It should be remembered that, as well as being a pathway for contaminants to be carried from one site to another, watercourses in the form of controlled waters are also potential receptors under Part IIA of the Environmental Protection Act 1990.

Fig. 8.3 Dredgings from the canal running through this city centre residential and office development were contaminated with heavy metals, the legacy of 200 years of industrial activity.

8.3 The Land Use Categories – a brief description

Brief descriptions of how contamination may be (or has been) caused in different industries are contained in this section of *Previously Developed Land*. Industrial activities have been combined together into 39 different classes but the reader should appreciate that these classes relate to several hundred different types of industries and manufacturing processes.

The source material for many of the 39 Land Use Categories has been the Industry Profiles that were published by the Department of the

Environment in 1995 and 1996, plus additional research by the author. The Industry Profiles were published in 48 volumes and the present work is not intended to be a substitute for the profiles but should be regarded as a summary of the types of activities which may have led to contamination, the nature of that contamination and where it may be located on the site. Readers are therefore encouraged to use *Previously Developed Land* as a 'ready reference' and to refer to the Industry Profiles for more detailed information, although it should be noted that many of the profiles are currently out of print. Tables 2.3 and 2.4 of Contaminated Land Report 8 (CLR 8) (Environment Agency, 2002b) are also based on the Industry Profiles and provide a useful checklist as to which contaminants may be associated with specific industrial uses.

The author wishes to acknowledge, with grateful thanks, the contribution to widening understanding about industrial activities and contamination that was made by the individuals and consultants who prepared the original Industry Profiles. In an attempt to further widen that understanding, many of the brief descriptions in this part of *Previously Developed Land* have been followed by short case studies, which contain many different lessons. These lessons have been drawn from the author's own experience and from the experiences of several leading site investigation and remediation specialists. The author would especially like to acknowledge the contributions made by the following people:

Dr Tom Henman at Enviros
Mike Smith at BAE SYSTEMS Property and Environmental Services
Mike Summerskill at SEnSE Associates LLP

Each of the Land Use Category descriptions starts with a list of the main groups of contaminants that may be encountered but it should be noted that there may be many sub-groups or individual contaminants within each group. The relevant DoE Industry Profiles are listed at the end of each description and a full list of the profiles is included at the back of this book.

8.4 Summary

Readers should bear in mind that, for many old industrial sites, different types of activities may have been carried on simultaneously, or may have succeeded each other. Interaction between the residues from these different activities may have resulted in the nature and concentration of contaminants being changed.

The handling and storage of raw materials, production processes and the disposal of wastes can all result in contamination. Whilst contamination from these activities may well be confined to the manufacturing site itself, contaminants may also be carried further afield through the air and ground or surface waters. Although the 39 land-use categories described in this part of *Previously Developed Land* encompass a considerable number of industries, they cannot hope to cover all types of industrial activities, let alone all possible contaminants.

It is therefore the responsibility of the site investigator, developer or valuer to consider all possible means by which contamination may have occurred and to investigate the site accordingly. The fact that premises have been used for one or more of the industrial activities described in this book does not necessarily mean that the site is, or has been, contaminated. Processes differed between manufacturing companies, as too did storage and waste disposal practices, and they especially changed over time. The brief descriptions and case studies are therefore intended for guidance purposes only and to pass on experiences gained from dealing with a wide variety of sites.

Industrial Activities

Contaminants, processes and case studies

AIRPORTS AND SIMILAR USES

Contaminants: *Inorganic compounds; Acids/alkalis; Asbestos; Solvents; Herbicides; Polychlorinated biphenyls (PCBs); Fuels; De-icing agents; Fire-fighting chemicals.*

Contamination at airports and similar facilities may arise through a number of key activities, including fuelling, de-icing, aircraft servicing/maintenance and other operations such as fire control. Other activities such as catering, car parking and freight handling may also give rise to contamination.

Fuels comprise aviation kerosene for jet engines, aviation gasoline for piston engines, and diesel for road vehicles. Fuel oil may also be used for space heating of airport buildings. The capacity of fuel storage facilities in airports may vary considerably and distribution within the airport may be by tanker or by underground pipeline. Filters control the quality of fuel and any unsatisfactory fuel may be transferred to a holding tank prior to being re-processed. Sub-surface drainage passes through interceptors and then to the main airport drainage system.

De-icing operations utilise chemicals including glycol, urea and acetate-based formulations. These are applied both to aircraft and runways. Urea granules may be stored in a hopper on site but this material is being replaced by calcium and magnesium acetate based products.

Servicing and maintenance facilities take place either inside hangar buildings or on the airport aprons. Wastewater effluent, from cleaning aircraft fuselages, is collected in large catch-pits, which are ventilated to the atmosphere to allow the evaporation of volatile organic compounds. Other potential contaminants include 'aluminium etch' primer paint, solvent based cleaning fluids, hydraulic oils and adhesives. Solid wastes may include Kevlar materials, metals and fibre glass.

Fire-fighting chemicals include foam concentrates (fluoro-protein chemicals), which are stored in 25 and 200 litre drums, dry powders and halon gases. These are used to extinguish practice fires for training purposes, as well as for emergencies, with run-off being directed towards the airport's interceptor system.

Soil contamination within airports is likely to be in the vicinity of drainage systems, including interceptors and soakaways, workshops or aircraft servicing aprons, fuel storage and distribution facilities, waste disposal facilities (including foodstuffs and sanitary disposal from aircraft holding tanks), building operations and leaks from electrical sub-stations.

Industry Profile: Airports

Undertaking remediation whilst minimising disruption

Probably the most likely cause of contamination at airports is leakage from underground fuel storage tanks, followed by accidental spills of fuels. Left undetected, plumes of hydrocarbon contamination can extend over considerable distances, carried in the direction of groundwater flows. Remediation without causing major disruption to air traffic, as well as to surrounding properties, then becomes a major problem. Fuel contamination at New York's J. F. Kennedy Airport was tackled using horizontal drilling. See Syms, 1997a, pp. 114–117 for more details about this technique.

ANIMAL SLAUGHTERING AND BY-PRODUCTS

Contaminants: *Acids and alkalis; Organic compounds; Pathogens; Metals and metalloid compounds; Inorganic compounds.*

This group covers a wide range of activities, all involving the slaughtering and processing of animals. It includes slaughterhouses and knacker's yards, hide and skin processing plants (e.g. tanneries and fellmongers), rendering plants for both edible and inedible products, glue and gelatine works and meat processing plants. Also included are pet food manufacturers. The locations where these activities are carried out have significantly reduced in number over the second half of the 20th century and the identification of historic contamination may therefore be rather more relevant than identifying existing sources.

Liquid blood and fat wastes may carry pathogens and have the greatest potential for the migration of contamination away from controlled areas in slaughterhouses and similar operations. Soil contamination may arise through leaking tanks, pipework or storage chambers, as well as through joints in concrete floor slabs and through permeable areas inside and outside buildings. Contamination of surface water may occur through the spillage of liquid wastes and, depending upon site-specific circumstances, groundwater may also be affected.

Metallic wastes have the potential to cause contamination in premises used as tanneries or fellmongers and are likely to be associated with sludge tanks, drying beds or open storage areas. Trivalent chromium and arsenic are common contaminants. Insecticides may have been used in some premises, especially fellmongers, and a common biocide is lindane, used as a sheep dip until it was banned in 1985. Lindane may be adsorbed onto soil particles and is biodegradable to some extent, with 75–100% disappearing over the course of 3–10 years.

Solvents, used for cleaning and in the tanning process, and fuel oils may be found in soils on animal processing sites. Small amounts of solvents can cause contamination of groundwater and may result in groundwater being rendered unfit as a drinking water supply, see *Cambridge Water Company* v. *Eastern Counties Leather Ltd*.

Industry Profile: Animal and animal products processing works

Setting appropriate laboratory detection limits – Abattoir
A house builder wished to buy land on the edge of a village, related to a previous recent use as an abattoir. The geology was a mixture of clays and sands, with alluvium in the valley floor below the site. The stream in the valley floor was the main Receptor at risk. Apart from property layout details, not much information was held on the site's geology and history.

The builder employed his 'usual' structural engineer to carry out a geo-environmental study. That investigation found buried blood tanks and an old diesel spill, in defined locations. The engineer had also tested for a range of pesticides, given the former animal use (sheep dips). One test showed minor levels of Dieldrin[1] and some other compounds. The value of the test was about five times the baseline detection of the laboratory used, but still within the 'inert' categorisation threshold for soil disposal. Nevertheless, the estimate for pesticide-affected soil removal was said to be in excess of £100 000.

The sale was held up for four months whilst the seller had his own detailed SI survey done, and the purchaser employed an environmental consultant. Neither of these further investigations replicated the single finding, nor did either consultant find evidence of other contaminants such as pathogens (tested for the first time). Indeed, the detection limit of the two other laboratories used was ten times higher than the first one, and 'industry standard'. Thus the level of Dieldrin first detected (whether a 'false positive' or not) would have been below 'normal' detection, and 'intervention' level, in both cases. The sale eventually proceeded, with the seller bearing the extra investigation and testing costs of around £10 000 each in a reduced land price.

[1] A highly toxic chlorinated hydrocarbon used as an insecticide.

ASBESTOS MANUFACTURE AND USE

Contaminants: *Asbestos; Metals; Inorganic compounds; Organic compounds; Fuels; PCBs.*

Asbestos may be found in a wide range of building materials, encountered on existing or former industrial premises. Therefore it is important to recognise that asbestos, as a contaminant, may be found in many locations besides premises used for its manufacture or processing. All forms of asbestos have good resistance to heat, electricity and chemicals. They also have high tensile strength and it is for these reasons that asbestos fibres have been used in the manufacture of many products, such as sheet materials, slates and pipes, insulation and lagging, sprayed coatings and paints, flooring materials, reinforced plastics, mastics and sealants.

The health hazards associated with asbestos vary according to the type of asbestos used, *chrysotile* (white asbestos), *crocidolite* (blue asbestos), or *amosite* (brown asbestos), the product in which it has been used and whether the fibres are encapsulated, and the location of use. All three types of asbestos were used in the manufacture of asbestos cement, which commenced in the United Kingdom in about 1900, and *chrysotile* is still in use today.

Asbestos was used in the manufacture of insulation boards but its use was discontinued in the UK in 1980. Both *crocidolite* and *amosite* were used in the manufacture of insulation materials, ropes, yarns, quilts and pre-formed insulation covers but were gradually replaced between the 1950s and 1970s with other products, including fibreglass, vermiculite and mineral fibres. Asbestos laggings and sprayed coatings were used in many factories, hospitals and other public buildings. Many such buildings may now be coming up for redevelopment or refurbishment and, as the potential for release of asbestos fibres increases with age, care should be taken to determine whether or not any asbestos contamination risk exists.

The main manufacturing period for asbestos products pre-dated the earliest effective waste legislation, which was the Control of Pollution Act 1974, and the disposal of wastes would generally have been uncontrolled, with inadequate documentation. Tanks may have been used on site for storing suspended asbestos cement waste prior to re-use. On some sites slurry ponds may have existed to allow wastes to settle out prior to disposal. Solid wastes, off-cuts, broken sheets and sub-standard materials were often sold for use as hardcore by builders and within factory sites.

Other wastes from the manufacture of asbestos products include metallic pigments, from the spray painting of insulation boards, natural and synthetic rubbers from the production of moulded materials, powders and liquid resins from the manufacture of friction materials, and a variety of solvents, commonly used for thinning resins and oils, cleaning equipment and tanks.

Industry Profile: Asbestos manufacturing works

Specialist contractors required

When a former asbestos products manufacturing plant was being redeveloped in the 1980s it was found that, even though all the manufacturing equipment had been removed and the premises thoroughly cleaned, the brick walls of the buildings were impregnated with asbestos. As a result specialist contractors had to be employed to decontaminate the site.

In use, asbestos can be found in many situations where redevelopment is contemplated, from cladding on pipes, use as sprayed on insulation to walls and other structures, to asbestos cement sheeting. The services of specialist contractors will be required to remove the asbestos and the contractor should certify all works. Although encapsulated asbestos cement sheeting and similar asbestos products may not be perceived as being particularly dangerous, care needs to be taken during demolition and removal, so as to ensure that broken pieces are not left behind to contaminate the soil.

CONCRETE, CERAMICS, CEMENT AND PLASTER WORKS

Contaminants: *Metals, metalloids and their compounds; Acids/bases; Asbestos; Solvents; Polychlorinated biphenyls (PCBs); Hydrocarbon (fuels).*

This group includes a number of different manufacturing processes, from bricks to tableware, cements to industrial ceramics. The principal raw materials used in the manufacture of these products (clays, shales, gypsum, lime) may not themselves be contaminative but contamination may arise through the use of additives and from by-products or wastes. Ancillary activities, such as vehicle workshops and plant maintenance areas, may also result in contamination.

Contamination may have arisen through leaks or spillages of oils and lubricants in non-bunded storage areas, or through loading or transfer operations. Wastes in the form of bricks, broken ceramic, kiln ash and workshop wastes may have been disposed of on-site, for example to fill hollows in the ground or to create hardstandings for vehicles. On-site disposal may also have included surplus metallic pigments from ceramic glazes. Some works may have had their own electrical transformers, which could have resulted in leaks of PCBs. Many of the process wastes may be in dust form, which could be significant when considering the potential of contaminants to migrate from the site.

Industry Profile: Ceramics, cement and asphalt manufacturing works

DISINFECTANTS MANUFACTURE

Contaminants: *Metals, metalloids and their compounds; Acids and alkalis; Asbestos; Solvents; Coal tar components including phenols, Polycyclic aromatic hydro-carbons (PAHs); Fuel oils and coal; Waste organic compounds (can be chlorinated); Poly-chlorinated biphenyls (PCBs); Dioxins.*

Disinfectants are substances that kill, or inhibit the growth of, harmful microorganisms. They contain an active ingredient, a biocide, to which other ingredients may be added and are supplied in liquid, solid or gas forms. In the general sense of the term, disinfectants are made for application to inanimate objects, whereas antiseptics (also containing biocides) are medical agents for application to wounds.

The manufacture of disinfectants developed through the 20th century and is widely distributed, with major centres in South Wales, the North West, Yorkshire, the North East, and the South East. Active ingredients are normally manufactured by a division of a major chemical company, which may also produce the final product for sale to the consumer. Consumer products are also produced by a number of smaller companies that are supplied with ingredients by the chemical companies.

Raw materials are either transferred within production sites or are delivered by road, with bulk materials in tankers, to the manufacturer. The type of containers used will depend upon factors such as whether or not the materials involved are corrosive or flammable. On-site storage will generally be in purpose-built chemical storage warehouses or in secure areas with bunds or other containment but, historically, drums may have been stored in the open without any secondary containment. Bulk liquids may be stored in large tanks, above or below ground, and these together with associated pipeworks may have contributed to soil contamination. Transfer of drums of ingredients within sites, by fork lift trucks, was another potential source of accidental contamination but current good practice requires the use of drum clamps.

A number of different processes may be used in the production of disinfectants but, for the most part, these are likely to be 'batch' processes, as against continuous production. It is also likely that a variety of products, involving different ingredients, may be manufactured on the same site. The nature of contamination found on a site will therefore depend upon the types of ingredients used, the nature of processes employed and the duration of operations. Contaminants may, for example, include solvents, chlorinated phenols and their derivatives, mercury, PCBs and asbestos. In addition to production and storage areas, residues of contaminants may be found in drainage systems and soakaways, and in on-site landfills.

Industry Profile: Chemical works, disinfectants manufacturing works

DOCKYARDS AND WHARVES

Contaminants: *Will depend upon the types of cargoes handled but may include: metals and metalloids, sulphates, sulphides and cyanides* (metal ores, coal/coke and ancillary activities such as coal carbonisation); *diesel, petrol, mineral oils, phenols, aliphatic and aromatic hydrocarbons, dispersing agents* (petroleum products); *pesticides and preservatives* (timber products, leather and other animal products, foodstuffs); *refrigerants* (cold and chill stores); *solvents, detergents, paint residues, oils and hydraulic fluids* (maintenance operations); *PCBs* (electrical transformers).

The likelihood and nature of contamination will largely depend upon the type of handling systems and transfer methods employed. Also of importance is whether goods are, or have been, handled in bulk, loose or liquid forms, or whether the traffic is entirely containerised. Historically, open conveyors, hoppers, hoists, grab cranes and other mobile plant have handled dry bulk cargo. The nature of these operations created considerable potential for spillage and for wind-blown dusts. These methods have now largely been replaced by slurry systems involving pumped handling, thus reducing, but not entirely removing, the potential for contamination to occur. Pumped handling of liquids has also become increasingly common, replacing transfers of liquids in drums and tanks.

In many ports dedicated areas exist for the handling of raw materials and products relating to clearly identifiable local industries, such as coal, petroleum, timber and some agricultural products such as grain or sugar. The nature of these activities and the products handled may, however, have changed over time.

Transit sheds and other warehouses may have handled a wide range of products, for which little, if any, documentary evidence remains, and spillages or other potentially contaminative incidents may have gone unreported. Cold and chill stores utilise refrigerants, which may have leaked from refrigerating plants and pipeworks. Inadequate floor insulation in cold stores may also have caused damage to the underlying ground in the form of 'permafrost'.

Dock dredgings, and dredgings from inland waterways, may contain contaminants which originated from industries in the locality, or from many miles up the river or canal. These contaminants may include metallic pigments and other metals from deposited solids, which may be largely insoluble. Other contaminants in dredged material may include fuels, bulk fertilisers and other bulk chemicals, as well as ships' wastes.

Underlying peat and organic clays and silts may be sources of gases, such as methane or hydrogen sulphide. Gas generation and migration may occur through changes in the water regime and trapped gases may be released during the course of construction or civil engineering operations.

Industry Profile: Dockyards and dockland

Water as a pathway for contamination
During redevelopment of a site in central Manchester, adjacent to the Manchester and Ashton Canal, high concentrations of heavy metals were found in the soil. It transpired that the site had been used for the deposit of material dredged from the bed of the canal. This had become contaminated over a period of many years as a result of industrial effluent discharges and surface water run-off from the many industrial premises bordering the canal.

ELECTRICAL AND ELECTRONICS MANUFACTURE, INCLUDING SEMI-CONDUCTOR MANUFACTURING PLANTS

Contaminants: *Metals and metalloids; Inorganic compounds; Acids; Alkalis; Asbestos; Polychlorinated biphenyls (PCBs); Organic solvents; Halogenated compounds; Mineral oils; Effluent treatment chemicals/sludges.*

This group covers an extremely diverse range of equipment, products and manufacturing processes. Electronic engineering is the manufacture of equipment that operates on very low current, whilst electrical engineering deals with high power equipment. Common processes include the assembly of printed circuit boards, manufacture of electronic components, transformers, batteries and lighting equipment, and the production of insulated wire and cable.

The industry has expanded rapidly, especially since the 1970s, with its main centre being in South-East England and other important centres in the North West, West Midlands, the North East and in Scotland. Most facilities are fairly small, employing fewer than 25 people, but there is a small percentage of very large plants.

The raw materials used in the industry vary according to the type of product but commonly include, conductors (e.g. copper or copper alloys, carbon and titanium) and other metals such as iron and precious metals, semi-conductors (e.g. elemental germanium and gallium arsenide) and insulators, which may be solid, gas or liquid. Polychlorinated biphenyls (PCBs) have been used since the 1940s in the electrical industry, mainly as a dielectric (non-conducting) fluid in transformers and capacitors for alternating current (a.c.) electrical use. The manufacture and use of PCBs is restricted by legislation in the United Kingdom but the use of PCBs in certain existing equipment was exempted from the ban. However, concerns over the effect of PCBs on North Sea mammals has resulted in a multi-national agreement to phase out and destroy remaining identifiable PCBs by 1999.

Solid and sludge wastes from the various manufacturing processes employed in the industry include absorbent materials used for cleaning and mopping up purposes, copper and plastic residues from drilling and deburring circuit boards, dusts containing metal particles, metal off-cuts and swarf. Other solid wastes may include scrap, or sub-specification assemblies and circuit boards, broken glass, wire, and packaging materials. Liquid wastes may include PCB contaminated oil from transformers, spent pickling and etching acids, contaminated rinse waters and electrolyte from batteries.

Waste handling and disposal practices have improved significantly in recent years but ground contamination may have occurred in production and waste disposal areas. Site-specific conditions, including local geology and hydrogeology will be important in assessing the persistence and propensity of contaminants to migrate.

Industry Profile: Engineering works, electrical and electronic equipment manufacturing works (including works manufacturing equipment containing PCBs)

ELECTRICITY GENERATING (excluding nuclear power stations)

Contaminants: *Metals, metalloids and their compounds; Coal; Fuel oils; Lubricating oils; Water treatment chemicals; Asbestos; Timber treatment chemicals; Solvents; PCBs and other transformer oils.*

Until the 1960s, with the exception of a few nuclear and hydro-electric power plants, all electricity generating stations in the United Kingdom burned coal. Large coal fired power stations, feeding the National Grid, were often located in coal mining areas. Other coal fired generating stations existed to serve specific industries or industrial areas and were located adjacent to the areas they served. The 1960s and 1970s saw the development of oil-fired power stations (often in coastal areas close to oil refineries) and a few coal fired plants were partially converted to burn oil. Between 1972 and 1984 a total of 107 power stations were closed and the recent trend, in developing new electricity generating stations, has been towards Combined Cycle Gas Turbines (CCGT), using natural gas for fuel.

In coal fired power stations, coal is reduced to a powder, for maximum combustion, and blown into the combustion chamber of a boiler where it is burnt as a suspension in air and combustion gases, with pressure being maintained by the incoming air. Water circulating in tubes in the walls of the boiler is turned into steam, which is then piped under pressure into turbines where the expanding steam turns the rotors. The rotors turn electromagnetic generators to feed electricity into transformers and thence to the Grid system for distribution.

Boiler scale comprising iron oxide and possibly copper oxide accumulates on the internal surfaces of the boiler, inhibiting heat transfer and requiring periodic cleaning. Chemicals used in the cleaning process include citric acid, hydrochloric acid, ammonium bifluoride and sodium bromate. Organic based inhibitors are also used to ensure that the oxides, rather than the metal of the boiler tubes, are dissolved. After cleaning the metal surfaces are 'passivated' using hydrazine and ammonia.

Ammonia is added to the 'closed circuit' water/steam system of the boiler/turbine unit in order to maintain alkalinity and hydrazine is added to remove oxygen in the water feed to the boiler water. Other chemicals, which may be added to the boiler water, include sodium hydroxide or sodium phosphate.

Wastes from electricity generating stations include Furnace Bottom Ash (FBA) and Pulverised Fuel Ash (PFA). FBA, which accounts for about 20% of the total ash, contains larger agglomerated particles of ash or clinker, which fall to the base of the boiler, whilst the lighter particles of PFA are collected by electrostatic precipitators. Both FBA and PFA comprise around 80% alumina-silicates, plus calcium, potassium, sodium and magnesium (6–15% of the total) and small percentages of heavy metals. Other potentially contaminative materials or wastes include fuel oil, asbestos, PCBs, other oils and solvents.

Industry Profile: Power stations (excluding nuclear power stations)

All may not be what it seems from the plans

A desk study of plans for a demolished coal fired power station in the North West provided a good indication as to where the buildings had been situated and where the main foundations and other underground structures were likely to be found. A site walkover revealed that all structures had been demolished and the site surface consisted of demolition rubble, albeit somewhat overgrown. What the plans did not reveal was that under the buildings were vast tanks for cooling water, extending many metres into the ground, and these had been filled with the demolition materials, including asbestos and other contaminants, from the power station.

The true situation was discovered when the contractor started on site, intending to excavate the foundations, only to find that the loosely packed fill had to be removed before redevelopment could commence. The increased cost contributed to the bankruptcy of the developer.

ENGINEERING (heavy and general)

Contaminants: *Metals, metalloids and their compounds; Acids and alkalis; Organic compounds; Polychlorinated biphenyls (PCBs); Asbestos; Inorganic compounds; Explosives and associated products; Radioactive sources; Organic chemicals inc. fuel oils, oils and lubricants; Cyanides; Mineral acids; Organic solvents/thinners; Biocides; Epoxies/vinyls; Oily sludges; Coal/ash.*

This group covers a wide range of industrial activities linked by a common thread: they are all involved with converting previously processed raw materials into finished products. They are not processors of natural raw materials, nor do they produce synthetics. Equally, they are not just assemblers of pre-formed parts, although in many cases a significant part of the end value of the product will be purchased from outside suppliers or sub-contractors. The activities encompassed within each of these industries will, almost certainly, extend from initial design to production of the completed product.

Railway works, involved with the manufacture and repair of locomotives, carriages and rolling stock, numbered over 900 in 1935, of which the railway companies owned 550. At the time of rail nationalisation in 1948, the number of major works building new locomotives or undertaking heavy repairs or rebuilding totalled 26. The principal raw materials used in railway engineering are steel, aluminium, timber, insulating materials (which historically included asbestos), plastics, fibreglass and various prefabricated components. Other materials include diesel oil, degreasing agents, paints, solvents and lubricants.

At the height of the Second World War some 22.15 million square feet (2.06 million square metres) of manufacturing floor space in the United Kingdom was devoted to aircraft production and related industries. Since that time the industry has significantly reduced in terms of both numbers of companies, through mergers and closures, and in production capacity. Early aircraft production relied heavily on the use of wood or bamboo frameworks, held together with piano wire or steel cables, covered with cotton or linen, impregnated with dope. Cellulose nitrate was the most commonly used doping mixture and was highly flammable, being closely related to the explosive guncotton. Modern aircraft production uses aluminium alloys and other metals, such as high tensile steel, nickel and titanium alloys. Copper, brass and other specialist alloys are used in mechanical and hydraulic systems. Reinforced plastic, fibreglass and carbon fibre reinforced materials such as Kevlar are used extensively in aircraft production. Other materials include primers, lacquers and enamel paints, natural and synthetic rubbers, natural and synthetic oils, solvents, acids and alkalis for cleaning, asbestos in insulating materials and hydrocarbon based fuels.

The generic term 'mechanical engineering' includes light engineering, producing small products such as components for larger machines, tools, implements, locks etc., and heavy engineering, producing large scale equipment such as bridge members, structural steel and industrial machinery. Some plants may also include metals casting, machining and surface treatments. Some parts of the country have historical links with specific types of engineering, e.g. Birmingham (small arms and cars), Coventry (bicycles then cars), Clydebank (sewing machines) and Lincoln (agricultural machinery). Whilst some of these industries have undergone significant reduction over the last two decades, the importance of these locations remains, especially when considering the redevelopment of previously used land. Materials used include ferrous and non-ferrous metals and alloys, plastics, lubricants (some of which may contain low concentrations of PAHs), surface treatment reagents, coating materials (e.g. red lead, zinc chromate, zinc oxide), paints, adhesives and other bonding materials, fuel oils, detergents and cleaning compounds.

The shipbuilding industry dates back many centuries, to the days of wooden ships, then iron through to steel and, in consequence, shipbuilding and repairing sites may have undergone several stages of development. Modern steel shipbuilding involves the fabrication of a complex steel structure, into which a wide range of equipment is fixed, with around 60% of the value being purchased from outside suppliers in the form of materials and services. Materials include steel and other metals, wood, plastics, insulation, solvents, finishes (primer paints, varnishes, antifouling etc.). Modern antifouling treatments are limited in terms of their active ingredients, which have to be approved for use under the Control of Pesticides Regulations 1986; however, shipbuilding sites may be contaminated with the residues from older antifoulings which contained metals such as lead, copper, tin, arsenic and mercury.

Vehicle production (cars and commercials) in the United Kingdom has declined since its peak in the mid-1960s and many well known marques have disappeared. Consequently production capacity has reduced and plants have closed, with many being redeveloped for other uses. Modern vehicle production is mainly an assembly process marrying together body parts with other components and fittings. Major components, such as engines and transmissions, are either bought in from specialist suppliers or from dedicated plants owned by the car production company. Pressed steel sheet is the main component of body parts, although aluminium is used by some manufacturers. Other materials include welding and soldering alloys, paints, solvents, sealants and waxes, fuel oils, transmission and engine oils, plastics (for upholstery and some body parts), glass, electrical wiring and electronic components.

Waste management procedures within engineering vary considerably, with some firms/industries being far more conscientious than others in terms of waste minimisation and recycling. Whilst today engineering plants may recycle metals, especially expensive alloys, and collect solvents and other liquids for off-site disposal or specialist reprocessing, this may not have been the case in the past.

The on-site disposal of waste by burying, discharge to sewers or soakaways, or simply depositing on the ground, was prevalent in engineering industries. Therefore contamination is likely to be present on older sites. Some of the contaminants, such as solvents and oils, may be highly mobile. Asbestos may be present, having been used for pipe and plant lagging, fireproofing or roofing of buildings. Some asbestos dust may have been produced by brake lining machinery. PCBs may be present from dielectric fluids coming into contact with the ground during refilling or dismantling of electrical equipment.

Industry Profile: Engineering works

- railway engineering works
- aircraft manufacturing works
- mechanical engineering and ordnance works
- shipbuilding, repair and shipbreaking (including naval shipyards)
- vehicle manufacturing works

Anecdotal evidence – Engineering works

Part of a much larger site undergoing redevelopment, the whole site had been the subject of a comprehensive Phase 1 study. This had then informed a method statement for the investigation and assessment of ground and groundwater contamination.

In light of historical plans the engineering works building was the subject of a targeted and non-targeted investigation providing an exploratory hole spacing of at least 25 m. No significant concentrations of organic or inorganic contaminants were recorded in the Made Ground or the underlying natural Superficial Deposits.

After demolition works on the site had commenced a former employee contacted the local authority about an alleged cyanide drum burial area. The drum burial area was apparently a structure that was filled in when a larger building was constructed over it in the 1960s. Although the current building had been extended there were no historical plans of former below-ground structures.

The area, which fell between the 25 m exploratory grid, was investigated via a trench across the suspected area and a former structure encountered. Brick and concrete lined, the structure was approximately 15 m long, varied between 1.5 and 2 m wide and was 1.8 m deep. No drums were present but the structure had been infilled with brick and concrete rubble. Yellow staining to some of the infill material suggested elevated chromium, not cyanide, concentrations. Laboratory analysis revealed significant concentrations of total and hexavalent chromium (4100 and 2600 mg/kg respectively).

Although the volumes to be dealt with were small the importance of anecdotal information should not be overlooked.

EXPLOSIVES INDUSTRY, INCLUDING FIREWORKS MANUFACTURE

Contaminants: *Explosive materials; Acidic effluent and explosive residues; Mineral acids; Non-chlorinated organic solvents/compounds; Chlorinated organic cleaning solvents; Inorganic salts; Inorganic compounds; Calcium as lime; Metals and metal salts; Non-metals, e.g. sulphur; Asbestos; Fuel oil; PCBs.*

Explosive materials are chemicals or mixtures of chemicals which, when suitably initiated, decompose with the rapid formation of hot gases. Explosive materials may be solid, liquid or gelatinous substances. Methods of initiation may be mechanical (e.g. impact or friction), heat induced (e.g. sparks or open flame), detonating shock (e.g. a blasting cap), or they may be caused by electrical or microwave stimuli.

Explosives manufacture in the United Kingdom has developed since the 16th century, therefore many sites distributed across the entire country may have either manufactured explosives or have handled explosive materials. A variety of raw materials are used in the manufacture of explosives, some of which are required in bulk and others in smaller quantities. Bulk materials include mineral acids, organic solvents and polymers. Smaller quantity materials include aluminium powder, sodium nitrate, sodium perchlorate and guar gum. Historically, the majority of raw materials were delivered by water or rail but these methods have largely been superseded by road vehicles and bulk tankers.

The main reagents in military high explosives are mineral acids (nitric and sulphuric acids), organic solvents (e.g. acetone) and organic compounds (e.g. hexamine and toluene). For commercial explosives the most important raw material is ammonium nitrate, which is combined with fuels such as liquid hydrocarbons and cellulosic materials. Nitrocellulose (NC) and nitroglycerine (NG) form the basic raw materials for propellants and are manufactured by adding mineral acids to cellulose and glycerine respectively. Metal powders (e.g. magnesium and aluminium), non-metals (e.g. boron, phosphorous), charcoal, sulphur, antimony sulphide and organic materials are the main constituents of pyrotechnics. Paraffin and ammonium phosphate are used in the production of matches and white phosphorous was used up to the 1930s.

High explosives manufacture produces significant volumes of liquid effluents containing acids, organic liquids and water, as well as small quantities of dissolved or suspended explosive. On-site recycling is commonly practised but this did not always apply in the past. Settling is employed in order to remove recyclable solid materials from wastewaters that may, historically, have been allowed to discharge directly into surface waters. Solid wastes include reject explosives, including propellants, explosives in settling tanks, floor sweepings and cleaning materials. Disposal of wastes may have been by on-site burning and this method is used today for the remediation of explosive-contaminated land.

Industry Profile: Chemical works, explosives, propellants and pyrotechnics manufacturing works

FERTILISER MANUFACTURE

Contaminants: *Metal and metallic compounds; Inorganic compounds; Acids/alkalis; Organic compounds; Asbestos.*

Fertilisers are inorganic or organic based nutrients that promote plant growth. They include naturally occurring materials, such as farmyard manure, and manufactured materials produced to specific formulations. Commercially produced fertilisers may be organic or inorganic.

The materials used in organic fertilisers include ground bonemeal, fishmeal, compost and manure. Inorganic fertilisers, which are produced by both batch and continuous methods, include nitrogen, phosphate, potassium and compound fertilisers. The nature of finished fertiliser products, and their quantity, depends upon the nature of the market for which they are intended. They may be supplied in bulk, for example by tanker, in bags, drums or kegs.

Waste materials will vary according to the type of fertilisers being manufactured and waste disposal practices at individual sites may have varied considerably over time. Today, wastes are generally collected and stored in properly designed compounds, with bunds where appropriate, and disposed of off-site by specialist contractors. Historically, wastes may have been disposed of by burying on-site, or to local landfills. Wastewaters are treated on-site but underground pipeworks may contain untreated residues. Some contaminants may contain traces of heavy metals and other wastes may include PCBs (from on-site sub-stations) and asbestos from demolished or altered buildings. There is also a small, but significant, possibility that spores of the anthrax bacterium *Bacillus anthracis* may be present, having been brought onto the site in imported bonemeal.

Industry Profile: Chemical works, fertiliser manufacturing works

FILM AND PHOTOGRAPHIC PROCESSING

Contaminants: *Metals and metalloids; Inorganic ions; Acids/ alkalis; Asbestos; Organics inc. hydroquinone, hydro-carbon fuel; Polychlorinated biphenyls (PCBs).*

The sensitivity to light of certain metallic salts was first used for photographic purposes in the 19th century and they are still used today. Modern photographic films are coated with gelatin layers containing grains of silver halide salts (e.g. silver chloride, silver bromide) to form emulsions, with chemical and spectral sensitisers to increase the sensitivity to light. When exposed to light, the silver halide molecules are reduced and converted into metallic silver and bromine or chlorine gas. The gas is lost and the silver atoms associate with the photographic emulsion to form a latent image. Colour photographic materials comprise superimposed layers that are sensitive to blue, green and red spectral regions. Photographic processing is the converting of the latent images into pictures, through three stages: developing, fixing and washing.

In addition to the metals and chemicals used in the production of film, other chemicals are employed in the processing stages. These include alkalis (e.g. sodium hydroxide), accelerators (e.g. acetic acid, cinnamic acid disulphide), preservatives (e.g. potassium sulphite, ascorbic acid), restrainers and anti-foggants (e.g. benzotriazole, potassium bromide), wetting agents and hardening agents.

Liquid wastes include spent chemical solutions which overflow from the processing machines through the replenishment process, together with substantial quantities of wash water. With the exception of very large commercial processing plants, discharge of liquid wastes is to the foul sewer without prior treatment. Recovery of silver is undertaken by feeding spent fixing and bleach solutions into storage containers where the metal settles out. Solid wastes, such as scrap film, may have been disposed of on-site.

Industry Profile: Profile of miscellaneous industries; incorporating photographic processing industry

FINE CHEMICALS, INCLUDING DYESTUFFS AND PIGMENTS MANUFACTURING

Contaminants: *Metals and metalloid compounds; Inorganic compounds; Acids/bases; Alkalis; Asbestos; Organic compounds; Fuels; Polychlorinated biphenyls (PCBs); Effluent treatment chemicals.*

The historical development of fine chemicals in the United Kingdom is closely associated with the development of synthetic dyes for the textile industry in the 19th century. Development of this industry largely occurred in the North West, due to the proximity of the textile industry and today around 80% of fine chemicals manufacture is located along the line of the M62, from Hull in the east to Runcorn and Liverpool in the west. The plants producing these chemicals are generally large scale, not dedicated to a single product but instead capable of producing up to a 100 separate products.

Dry and liquid (solvent) raw materials may be delivered to sites in a wide range of containers, from sacks and drums to road or rail tankers and even ships. Gases are also used and may be delivered in pressurised cylinders, tanks, or in bulk by road tanker.

Dyestuffs and pigments are synthesised from a fairly small number of raw materials, e.g. cyclic aromatic compounds (including benzene, toluene, xylenes, napthalene and aniline), many aliphatic organics and inorganics, such as sulphuric acid, chlorine, nitric acid and sodium nitrate. Batch processing is carried out in cast iron, mild steel, glass-lined or stainless steel reactors.

Wastes from the manufacturing process include wastewaters and other liquid effluents, solid wastes and solvents. Wastewaters are typically discharged to the public sewer after undergoing initial treatment on site, which may be limited to simple settlement and pH adjustment but may include treatments such as anaerobic digestion. Liquid wastes that are unsuited to disposal via the public sewer may be removed by tanker for off-site reprocessing, or for disposal at a suitably licensed facility. Solid wastes, such as spent activated carbon, may also be removed for off-site disposal to landfill (or for incineration), although it should be noted that some older plants might have on-site closed landfills or redundant sludge lagoons. Spent solvents are recovered for reprocessing either on-site or off-site.

Industry Profile: **Chemical works, fine chemicals manufacturing works; Chemical works, coatings** (paints and printing inks) **manufacturing works**

Even the best records may not be complete – Dyestuffs manufactory

The redevelopment of a former dyestuffs manufacturing plant, which had been in operation for around 80 years, involved extensive site remediation. This comprised the excavation and removal of contaminated soils to depths of between two and four metres across almost the entire site. The remediation work followed an extensive site investigation, with a mix of trial pits and boreholes on a 25-metre grid. Extensive contamination was identified and classified by type and concentrations. Estimated quantities indicated that the volume of contaminated soils within the site would be significantly greater than the volume of clean soil. As the intention was for the contaminated soils to be removed to off-site landfill the decision was taken, for project costing purposes, to assume that all soil to an average depth of three metres would have to be disposed of. Any 'clean' soil retained on site would then represent a saving in terms of the remediation cost. Floor slabs and other over-site concrete were to be crushed on site and, if certified free from contamination by the resident chemist, would be re-used as part of the new development.

The site investigation took place before all buildings on the site had been demolished and, whilst external concrete areas were broken open for trial pits or the sinking of boreholes, the floorslabs of buildings were generally left undisturbed. It was not until most of the buildings had been demolished and the site remediation work was underway that one particular floorslab, approximately one metre in thickness, was broken up. Removal of the concrete revealed buried nitrobenzene mixed with soil, approximately 1000 tonnes in volume, with a concentration in excess of 20%.

The company that had operated the site for more than 50 years had made all of its records available to the site investigation team and confirmed that it had never used nitrobenzene in any of its processes. Nitrobenzene is an oxidant, flammable when exposed to heat and flame, with a moderate risk of explosion. It is also a human poison and had seemingly lain undetected on the site since the building was constructed more than 50 years earlier. Removal of the contaminated soil and its disposal as a Special Waste more than used up any possible savings that might have accrued from the retention of clean soil on the site and resulted in a substantial six-figure sum of additional expenditure.

FOOD PROCESSING, INCLUDING BREWING AND MALTING, DISTILLING OF SPIRITS

Contaminants: *Acids and alkalis; Organic compounds; Pathogens; Metals and metalloid compounds; Inorganic compounds.*

This group covers a wide range of activities, some of which are included within other industry descriptions. The potential for contamination from food industry activities is diverse but is mostly related to the disposal of wastes.

Liquid animal fats and other oils, often in a semi-dissolved state in waste waters, may have been disposed of direct to the ground, where they can persist for many years, especially in silty clays. Although not necessarily of a seriously contaminative nature, the resultant odour, when the ground is opened up for redevelopment, can be highly offensive.

Sugars and molasses, both from refining processes and from the manufacture of jams and confectionery also may have been disposed of direct to the ground. These may have deleterious effects on foundations and underground services. Even where liquid wastes have been disposed of to foul sewers, contamination can occur through defects in the sewerage system.

Dairy products, such as cheeses, and breads or other yeast containing foodstuffs may have been disposed of by burying on-site, or to landfills. These can both contribute to landfill gas and produce contaminated leachate. Buried vegetable wastes have a considerable potential to produce landfill gas.

Industry Profile: None but see **Animal and animal products processing works**

Leaking drains and on-site disposal of wastes – Brewery site

Site investigations were undertaken on a major brewery site to support the development of a new packaging plant and to provide baseline information for the site's pollution and prevention control (PPC) permit application. The site had been operated since the late 19th century and potential sources of contamination identified were associated with former waste tips and coal storage yards and ongoing activities associated with the brewery, including chemical and oil storage and leakage from effluent drains. Site investigations identified the presence of localised hydrocarbon contamination, which required remediation to protect groundwater resources and to comply with planning conditions for the new development. This involved excavation of contaminated soils following decommissioning of bulk storage tanks. Leakage from effluent drains was also found to be a significant issue, as indicated by high biochemical oxygen demand (BOD) levels in shallow groundwater in one part of the site.

GARAGES, INCLUDING SALE OF AUTOMOTIVE FUEL, REPAIR OF CARS AND BIKES

Contaminants: *Metals and their compounds; Acids/alkalis; Asbestos; Organic compounds; Ethylene glycol; Polymerised glycol and ethers; Detergents.*

The storage, handling, leakage and disposal of raw materials and waste products from petrol filling stations and garage repair sites may all lead to soil contamination. Contamination due to overfilling, faulty pipes or caps was not uncommon in the past but much stricter controls are in force today, although accidents may still happen. Leaks may still occur from underground storage tanks and pipework; redundant tanks may also present problems unless they have been properly decommissioned.

The disposal of waste oil from vehicle servicing may have consisted of nothing more than pouring it down the drains or dumping it on open ground within the site or on adjacent land. Even when holding tanks have been used, pending collection of waste oils by a contractor, the areas around such tanks may be contaminated as the result of careless disposal or inadequate bunding. Waste materials, in the form of surplus paints, solvents and metals from repair facilities may have been disposed of on site. Car washes are another potential source of contamination, with waste water containing the chemicals used for cleaning.

Fuels such as petrol and diesel are highly mobile and have the potential to cause widespread contamination. This may result in free product forming a migratory plume away from the site; vapour may also diffuse into the soil and advance ahead of the free product. Lighter fractions of the fuel may float on the surface of groundwater, with heavier fractions migrating downwards through the water. Contaminants are less mobile in clay soils and those with a high organic content, which will adsorb the contaminants, are most mobile in coarse-grained sands and gravels.

Methyl tertiary butyl ether (MtBE) (an anti-knock additive to petrol) is at least ten times more soluble in water than the other constituents of petrol, its taste threshold is extremely low (10 µg/l) and it may taint potable water supplies at low concentrations. Other potential contaminants include, chlorinated hydrocarbons (used as degreasing solvents), heavy metals (additives in some oils and fuels) and battery acids.

Industry Profile: Road vehicle fuelling, service and repair

- garages and filling stations

On-site remediation

The site of a petrol filling station and motor vehicle workshop was to be redeveloped for housing. In a rural location, close to the centre of a Cotswolds village, road access to the site was difficult, as it was situated on a bend, adjacent to a narrow bridge over a river. Contamination of the site had occurred from leaking underground storage tanks and the workshop activities, resulting in hydrocarbon and heavy metals contamination. Water quality in the adjoining river was good and the site was on the boundary of a major aquifer.

Due to the access problems the number of vehicle movements had to be kept to a minimum and so removal of the contaminated soil was not possible. Therefore on-site bio-remediation was decided upon. The first stage of the remediation involved the removal of underground storage tanks, breaking up and crushing the over-site concrete for re-use. The contaminated soil was then removed and treated in windrows, where nutrients were added to speed up the process. The bio-remediation took six weeks, during which time the ground and surface waters were monitored for contamination. A small quantity of soils containing heavy metals was removed from the site although in the opinion of the environmental consultants this was unnecessary (see Syms & Knight, 2000, Chapter 9 for more details).

GAS WORKS, COKE WORKS, COAL CARBONISATION AND SIMILAR

Contaminants: *Ammonical liquors; Coal tars; Spent oxide; Foul lime; Metals; Coal dust.*

The coal carbonisation industry developed through three main branches: coke production (for the iron and steel industries), gas production (for industrial use, street lighting and later domestic use) and tar distillation. Coal gas was widely used for illumination in the United Kingdom from the early 19th century and by the early 1820s town gas works were established in many towns and cities.

Following the introduction of natural gas, from the North Sea and elsewhere, the production of coal gas ceased and the gas works became disused, with the possible exception of the gas holders which may have been retained in use for the storage of natural gas.

Ownership of many of the former gasworks sites was transferred to British Gas and in the late 1990s BG plc owned around 800–900 such sites, with varying degrees of contamination. Other disused sites were retained by local authorities and may have been redeveloped. The use of coke has declined, although it is still used in the metallurgical industries, and in 1995 there were only four remaining sites in the UK.

Coal carbonisation involves heating the coal, with a limited oxygen supply, which produces a gas stream from the ovens containing impure coal gas, ammoniacal liquid and tar. High temperature carbonisation was used to produce coal gas, which was subjected to a cleaning process (to remove ammonia, tar, hydrocarbon oils and sulphur) before distribution. Lower temperature carbonisation is still used for the manufacture of smokeless fuels, producing lower gas yields and a coal tar which contains a high proportion of the volatile components.

Waste products from the coal carbonisation process include ammoniacal liquors, coal tars, spent oxide (containing ferric ferrocyanide, used in the dyeing industry as Prussian Blue), foul lime (slaked lime which was used in the gas purification process), heavy metals and coal dust. Perhaps more than most other industries the potential for contamination from coal carbonisation sites will depend on the age of the plant, its period of operation and the nature of the processes undertaken. Contamination may have arisen through on-site disposal of wastes in landfill, or in lagoons, open storage of materials or wastes, from process buildings and tar pits, underground tanks and pipes. Other causes of historic contamination may have included the direct release of liquid wastes to ground or sewer.

Some distillates of coal tar, e.g. phenol, are water soluble and may migrate in solution through the soil to groundwater. Phenols may also migrate through plastic pipes. Other distillates, e.g. benzene and toluene, are mobile but are less soluble in water. Polycyclic aromatic hydrocarbons (PAHs) are less mobile but may migrate depending upon soil conditions. Free cyanide and sulphates may be mobilised by rainwater infiltration, with the potential to cause groundwater contamination. Sulphates can also attack concrete and cast iron. The movement of metals through the soil is retarded by the presence of clay minerals and organic matter.

Industry Profile: Gas works, coke works and other coal carbonisation plants

GLASS MANUFACTURE

Contaminants: *Metals and metalloids; Inorganic ions; Acids; Alkalis; Asbestos; Organic compounds inc. solvents, lubricating oils, Polychlorinated biphenyls (PCBs) and hydrocarbon fuels.*

Glass manufacturing techniques have altered little since the Middle Ages, with most changes relating to the additives used to make different types of glass, or the fuels used in the melting process. The principal raw materials are sand (quartz), soda ash (calcium carbonate), feldspar (a group of alumino-silicate minerals), limestone, recycled glass, anthracite (a reducing agent) and borax (sodium borate decahydrate). Minor ingredients include cerium, carbon, selenium, cobalt and various sulphates. Lead oxide is used in the manufacture of lead glass. Various other metals may be used in the manufacture of coloured and coated glasses. Acids, caustic solutions, solvents, lubricating and fuel oils are also used in the production and finishing processes.

Glass manufacture can be classified into three broad types: flat glass, glass containers and other glass products. Melting of the raw materials takes place, in either a pot or tank furnace, at temperatures of at least 1200°C after which the molten glass is shaped, moulded or floated according to the type of manufacturing process. After shaping, the glass is annealed or strengthened so as to reduce strain in the product. Finally the glass goes through one or more finishing processes, including cleaning, grinding, polishing, cutting, sandblasting and enamelling.

Waste materials include sub-standard, broken, mis-formed and overflow glass (cullet) that is recycled on-site. Other wastes include various oils, which may be contained in holding tanks prior to off-site disposal, and waste sludges from decoration or acid neutralisation processes. Waste process powders, containing high concentrations of selenium, cobalt or borate compounds, are sent to appropriately licensed disposal facilities.

Fuel oils, which are used in large quantities, and lubricating oils are the main potential contaminants. Some glassworks sites may have contained coal gasification plants and contaminants associated with that use may also be present. Metallic residues, from additives, may be present in locally significant concentrations. Solvent and acid contamination may have occurred in degreasing and washing areas.

Industry Profile: Profile of miscellaneous industries incorporating glass manufacturing works

IRON AND STEELWORKS

Contaminants: *Metals and metalloids; Inorganic compounds; Acids/Alkalis; Asbestos (roofing and lagging); Organic compounds.*

Steel manufacture developed as a large-scale industry in the United Kingdom after 1850, with large-scale works being built in areas close to extensive limestone deposits. Limestone is used, together with coke, in blast furnaces to smelt iron ore in order to produce iron. Steel is an alloy of iron obtained by refining iron to remove excess carbon and other elements, e.g. silicon and phosphorous.

Large-scale steel producing plants may contain six distinct processes: iron ore preparation; coke making; iron making; steel making; casting (ingot or continuous); rolling and finishing.

The principal wastes created by iron making are blast furnace slag, dry dust and wet solids from gas cleaning and refractory waste from the ladles and runners. Similar wastes and arising scrap material are produced from the steel making process. Metallic wastes also arise from casting and finishing operations. Scrap metal wastes are recycled to the steel furnaces. Pickling operations produce waste sulphuric and hydrochloric acids and sodium hydroxide.

The types of soil contamination found on an iron or steel works site will depend to a very large extent on the nature of the operations carried out and their duration. Contamination may be found in or near the process plant itself, in waste storage areas, wastewater drainage systems and wastewater treatment plants. Site drainage may have been to soakaways, thus allowing direct contamination of the soil. PCB contamination from transformers may also be present.

The potential for metal contamination to migrate within, and from, the site will depend to a significant extent upon the nature of the soil and its pH. Organic clay soils will slow the migration of metallic contaminants, whilst increased acidity will increase their solubility and hence their potential to leach into groundwater.

Industry Profile: Metal manufacturing, refining and finishing works

- iron and steelworks

Secure Containment Facility – Iron and Steelworks
Ravenscraig Steelworks was one of the largest industrial landholdings in Europe with a long history of coal working and iron and steel manufacturing. Closure of the works in 1992 left a legacy of environmental issues requiring remediation to facilitate redevelopment of the site and led to the formation of a site management team including environmental specialists to deal with site clearance, site investigation, remediation and water management. A key environmental issue involved the presence of significant quantities of contaminated soil materials and liquids including by-products from coke manufacture.

A waste management strategy for the site was developed including the construction of an engineered Secure Containment Facility (SCF) following agreement with regulatory authorities. Work carried out involved the construction, operation and restoration of the SCF and excavation and transport of contaminated materials to the SCF. The design for the SCF took into account final restoration landscape and future use to reflect the site's local setting and consistentency with subsequent development.

Development of the SCF also required planning consents and development of an environmental statement addressing landscape and visual aspects, potential nuisance issues during earthworks and potential effects on surface water and groundwater and soil gases.

LAUNDRIES AND DRY-CLEANING

Contaminants: *Organic compounds including solvents, Polychlorinated biphenyls (PCBs) and hydrocarbon fuels; Asbestos.*

Laundries are involved in the 'wet process' of cleaning garments and other textiles, whilst 'dry-cleaning' is a dry process of cleaning garments, which may include leathers and other 'non-textiles', through the use of organic solvents. Wet process laundries involve the use of significant quantities of water, together with detergents, bleaches, starches and other additives. Some sites may contain both laundry and dry-cleaning operations but today the majority of dry-cleaning operations are located in shop units in shopping areas.

The solvents in general use, for dry-cleaning, during the 20th century have included carbon tetrachloride (from the 1890s to the mid-20th century); trichlorethylene (from the 1920s); and, from the mid-1950s to the present day, perchloroethylene (tetrachloroethylene, tetrachloroethene). Chlorofluorocarbons (CFCs) have been used as dry-cleaning solvents (accounting for about 10% of the solvent used by the industry in 1990) but this use is being reduced in view of the effect of CFCs on atmospheric ozone depletion.

Most dry-cleaning operations are based around a cage, into which the garments are placed. The process generally comprises an initial solvent 'wash', followed by a spin to remove the solvent. The garments are then rinsed in distilled solvent, undergo a further spin and are dried in warm air. Small quantities of water, containing detergents, may be used to improve cleaning performance or to remove water-soluble stains. For the most part dry-cleaning machines are self-contained, fully automatic and incorporate solvent recycling.

Waste waters from both laundries and dry-cleaners are discharged to the public sewer and, at some larger sites, may undergo a degree of pre-treatment. Wastes produced by the dry-cleaning process consist of insoluble materials, such as extracted dirt/soil and filter powder; soluble fats and mineral oils; together with small quantities of spotting chemicals, water and solvent residues which have not been recycled. The most likely cause of contamination is the accidental spillage of solvents, although small quantities of other contaminants may accumulate in drainage systems.

Industry Profile: Profile of miscellaneous industries, incorporating dry-cleaners

METAL SMELTING AND REFINING, INCLUDING FURNACES AND FORGES, ELECTROPLATING, GALVANISING AND ANODISING

Contaminants: *Alkalis; Mineral acids; Organic acids; Oil; Organic solvents; Electroplating metals; Other metals inc. beryllium, aluminium, magnesium; Inorganic compounds inc. borates, cyanides, chlorides, sulphates, nitrates, phosphates; Asbestos; Polychlorinated biphenyls (PCBs).*

Iron is produced by the smelting of iron ore with oxides in a blast furnace to reduce the iron oxide to iron. The iron ore is also mixed with coke and limestone. These materials are tipped through the top of the blast furnace with heated air blown through nozzles at the bottom. The iron thus produced can be cast into solid pieces known as 'pig iron' or left in its molten state. Slag is produced as a waste product that can have a number of uses such as a building material or aggregate for road building. Other wastes produced include dry dust from gas cleaning, waste solids from gas cleaning and refractory waste.

The hot metal is then sent to the steel furnace where, following the production process, molten steel can be cast. Waste products include slag, scrap, dust and slurries and refractory materials.

Contamination may occur near storage areas, drainage systems and where leaks may have occurred from tanks and pipework. Transformers and other electrical equipment may contain PCBs, which have a low solubility in water and do not degrade. Works were traditionally located near ports and steel making was established in the Midlands and Sheffield.

The earlier forms of plating involved the fusing of silver plate onto copper ingots. This method emerged in the mid-1700s in Sheffield and Birmingham, with the first commercial electroplating works opening in Birmingham in the early 1800s. In the 1840s this method was further refined with the use of alkaline cyanide baths. A number of other metal plating processes subsequently became available including the application of gold and silver in thin layers, electroless deposition, hot-dip deposition, anodising etc.

Electroplating involves the deposition of a thin layer of metal, such as brass and bronze, by passing an electric current through an aqueous solution of the relevant metal salt. Contamination may result from leaks or spillages of process materials or wastes. In particular, contamination may occur around floor drains and tanks. The metal salts are water-soluble although the degree of solubility may be affected by soil and ground conditions. Cyanide may occur as free cyanide anion and as complex metal/cyanide anions. The slow breakdown of complex cyanides may act as a long-term source of free cyanide contamination to groundwater.

Hot-dip galvanising involves the immersion of iron and steel products into molten zinc. Prior to galvanising, the substrate is degreased and shot blasted or acid pickled. These items are used for large steel structures such as bridge components, marine pilings etc.

Anodising is an electrolytic treatment resulting in the oxidation of the metal surface to form a film of metal oxide.

Wastes in the form of liquid effluents, sludges and solids are produced. Liquids are usually discharged to the foul sewer following treatment, although untreated liquid effluents could, in the past, have been disposed of on-site or to soakaways. Sludges and solid wastes, such as tank sludges from process tanks and metal fragments, would usually be disposed of to landfill, frequently after treatment to immobilise heavy metals.

It has been estimated that there are in excess of 2000 electroplating and other metal finishing works in the United Kingdom. Although these are located throughout the UK, the highest concentrations occur in the Midlands, London and the South East, Lancashire and Yorkshire.

Industry Profile: Metal manufacturing, refining and finishing works: electroplating and other metal finishing works.
Metal manufacturing, refining and finishing works: iron and steelworks

Negative to positive land values – Metal Processing Works

A Phase I Land Quality assessment of a former metal processing works located in the Midlands revealed a number of potentially contaminated land uses including unlined former landfill sites. The property had been extensively filled and the Superficial Deposits and underlying economic mineral deposits had been extensively exploited. The Regulatory Authorities had also postulated that the site was affecting controlled waters.

A Phase II Land Quality Assessment was subsequently undertaken in order to confirm the geoenvironmental and geotechnical development abnormals associated with redeveloping the site for a mixture of end uses and to quantify the environmental liability associated with the site. The site investigations revealed limited areas requiring offsite removal of contaminated soil. This included a small area of potentially radioactive contaminated soil associated with imported scrap metal that had been used for smelting in the past. A plume of contaminated groundwater, enriched with trace metals, was

continued

(continued)

also identified. However, further work proved that although the plume was likely to be affecting controlled waters, concentrations of contaminants at the sensitive receptor were below guideline values.

Before the Phase I and Phase II Land Quality Assessments were undertaken, given the site's history the client feared that the site had a negative land value. Through a carefully phased programme of works to confirm contaminated land, geotechnical and liability issues, a positive land value has been realised. A remedial strategy has been developed and is awaiting regulatory approval.

MINING AND EXTRACTIVE INDUSTRIES

Contaminants: Lead, zinc, copper, tin, arsenic, cadmium.

Mining of coal was undertaken both underground, by the sinking of mine shafts, and by opencast methods. In both cases, large amounts of waste are produced and are usually deposited on-site and known as colliery spoil heaps. These spoil heaps are a source of contamination and, due to the amounts of coal contained therein, are a constant fire hazard. Temperatures of up to 1000°C have been recorded in burning heaps, thereby destroying all organic material. Once opencast mining ceases, legislation requires that the land is restored. Whilst such land can be reclaimed and built upon, it is more likely to be used for agriculture and leisure purposes. Records of coal mining are readily accessible, although it was not compulsory to maintain records until 1873, but documentation of other mining activities is frequently inadequate. Where there is any reason to suspect that mining activities may have occurred, careful inspection of Ordnance Survey maps may assist.

Other examples of mining are limestone in the West Midlands and salt extraction in Cheshire. Both of these can lead to extensive subsidence problems. Lead mining in the Peak District was by open shallow drafts, many of which are unidentified.

Pollution from abandoned coal and metal mines, together with spoil heaps, is a major problem where water is discharged or escapes into watercourses and the Environment Act 1995 provides new powers to deal with mines abandoned after 31 December 1999. The majority of the abandoned metal mines are in the south west of England with approximately 1700 being recorded as abandoned. Many of these sites are heavily contaminated by metals due to the method of extraction. However, arsenic can occur naturally with tin, and cadmium with lead and zinc, presenting hazards to vegetation and livestock. More watercourses are affected by contamination from abandoned metal mines than from coal mines.

Industry Profile: None

Failure to impose environmental conditions in planning permissions

In some parts of the UK, notably the South West, arsenic can occur naturally at concentrations that are well in excess of normally accepted levels for residential development. The natural concentrations can also be heightened by the tailings from long forgotten mining activities.

When granting planning permission for a small residential development in the grounds of a large house, the local planning authority neglected to impose any environmental conditions, such as those recommended in PPG 23, requiring a site investigation and remediation proposals. It did, however, impose a tree preservation order on the mature trees that formed the boundaries between the site and the road.

In accordance with the requirements of the National House-Building Council's Chapter 4.1, the developer undertook a site investigation, including laboratory analysis of soil samples, which revealed unacceptably high concentrations of arsenic. The developer, who was well used to dealing with local conditions, would normally have dealt with the problem by removing the topsoil and replacing it with clean imported soil but this would cause unacceptable damage to the tree roots. A number of compromise solutions were considered but none were acceptable to all three parties – developer, local authority and the NHBC. In the end it was decided that the only way forward was to remove the trees and the planning authority, having failed to impose any environmental conditions, found itself in the difficult position of being unable to enforce the tree preservation order and could only agree to their removal.

OIL REFINING, PETROCHEMICALS PRODUCTION AND STORAGE

Contaminants: *Metals, metalloids and their components; Inorganic compounds; Hydrocarbons; Other organic compounds; Asbestos; Polychlorinated biphenyls (PCBs).*

Crude oil is the basic raw material for all refinery processes and the practice of refining predates the petroleum industry, beginning in the 18th century with the simple distillation of raw materials. Crude oils are primarily complex mixtures of hydrocarbons, ranging from dissolved gases to compounds that are solid at ambient temperature. They also contain compounds with small amounts of oxygen, nitrogen and sulphur, as well as traces of metals such as vanadium, nickel and iron. There are marked differences in the composition of crude oils from different sources.

The principal products of oil refineries include refinery gas, liquid petroleum gas (LPG), naptha, motor spirit, aviation fuel, kerosene, gas oil/diesel, fuel oil, lubricants and waxes, bitumen and coke. The location of oil refineries is largely dictated by the proximity of deep-water ports and markets. In consequence, oil refining has been carried out at relatively few locations in the United Kingdom. The exploitation of North Sea oil since the mid-1970s did not change this situation, due to the ease with which the oil could still be shipped to the refineries.

Oil refineries are classified according to the complexity of their process operations:

Category 1 Topping/reforming is the simplest and provides basic crude distillation and catalytic reforming.
Category 2 The addition of a major cracking process.
Category 3 Adds lubricating oil manufacture to Category 2.
Category 4 Category 2 and/or 3 refineries plus a petrochemical plant.

Oil refining produces gaseous, liquid and solid waste streams. Gases may be flared off; wastewater containing free oil and solids are treated on site by various processes such as gravity separation, flotation and chemical coagulation. The treatment process yields free oil and/or sludges that have a high water content, requiring dewatering prior to disposal. The removal of impurities, using sodium hydroxide, produces a strong alkaline liquid waste containing phenols, cyanides and sulphur-containing compounds, frequently with very high concentrations.

The storage and distribution of oil-based products, held in tanks or flowing through pipelines, may have resulted in ground contamination through lack of integrity or accidental damage. Tanks for the storage of the more volatile petroleum products are subject to licensing under the Petroleum (Consolidation) Act 1928 (amended) and must be inspected for integrity on a programmed basis.

Hydrocarbon liquids released from the surface or from leaking underground storage tanks or pipelines will flow down through the ground under the influence of gravity. Lighter fractions will float on the surface of groundwater as 'free product', whilst insoluble fractions which are denser than water will sink through the groundwater until they encounter an impermeable barrier, at which point they will spread out horizontally and move in the direction of the groundwater flow.

Industry Profile: Oil refineries and bulk storage of crude oil and petroleum products

ORGANIC AND INORGANIC CHEMICALS PRODUCTION (not included elsewhere)

Contaminants: *Inorganic compounds; Alkalis; Asbestos: Organic compounds inc. vegetable oils, phenols, alcohols, glycols and derivatives; Polyclyclic aromatic hydrocarbons (PAHs), Polychlorinated biphenyls (PCBs); Fuels (e.g. oil, coal); Metals, metalloids and metal compounds; Acetylene and derivatives; Methane and derivatives; Ethylene and derivatives; Propylene and derivatives; Petroleum and derivatives; Benzene and derivatives; Pesticides and biocides.*

The bulk inorganic chemical industry is primarily concerned with the large-scale conversion of relatively simple raw materials into feedstocks for further processing. The main compounds produced by the industry are chlorine, sodium hydroxide, sodium carbonate, sulphuric acid, nitric acid, ammonia, phosphoric acid, ammonium nitrate, urea and calcium hydroxide (lime).

Most of the raw materials used by the inorganic chemicals industry arrive at the production plant in bulk quantities by road, rail or sea. Within the site itself bulk ores are typically transported by conveyor systems and liquids by pumping through distribution pipelines. Drums and kegs of smaller volume materials, such as reaction catalysts, may be moved by mechanical or hand trucks.

Waste products from chlor-alkali plants using the le Blanc process may have been disposed of in spoil heaps or in localised landfills within the vicinity of the plant. Sodium carbonate production, using the Solvay process, produced calcium chloride brines that were typically stored in large lagoons. Both of these waste disposal practices may have resulted in fairly widespread contamination, both horizontally and vertically through infiltration of rainwater, leaching and the run-off of liquid wastes.

Contamination from sulphuric acid works may have been caused by accidental releases of materials in the vicinity of the production plant, storage and transfer areas. In addition to the sulphuric acid itself, other contaminants may include heavy metals (associated with pyritic ores), ammonia and its derivatives (nitrates and nitrites) and cyanides. Nitrogen compounds may occur around ammonia and nitric acid production plants, with sulphur compounds possibly being encountered on sites operating before 1925. Phosphoric acid works may have resulted in contamination from the mineral fluorapatite, which may contain traces of uranium oxide. Fugitive fluoride emissions from process plant, particularly grinding operations, may have caused widespread low-level contamination extending beyond the site boundary.

Synthetic organic chemicals are manufactured from natural organic materials, such as petroleum, natural gas and coal that have undergone at least one chemical reaction, e.g. oxidation, hydrogenation or chlorination. The growth of the oil refining industry since the Second World War has been the main source of both aliphatic and aromatic primary organic chemicals, although coal based chemical production still does exist.

Modern organic chemical manufacturing works tend to be located within, or adjacent to, large oil refinery and petro-chemical complexes. The length of time over which the plant has operated is extremely important in the context of contamination, in particular on older sites through the storage and disposal of wastes. In the past little care was taken with the management of these materials, although waste management practices have improved through the self-regulation of the industry and as a result of legislation.

Waste materials include unreacted source materials, which are generally recycled back into the process unless this is uneconomic, and by-products and side-products, for which there may be little possibility of economic utilisation. Tars, filter cakes, precipitated compounds, spent catalysts and solvent containing residues are also produced. Spent scrubber solutions from washing process streams may be acidic or caustic. Containers used for the transport of materials to and within the site may also be contaminated.

Industry Profiles: Chemical works

- inorganic chemicals manufacturing works
- organic chemicals manufacturing works
- soap and detergent manufacturing works

Choice of remedial options – former chemical plant

Part of an industrial development site in East London was formerly occupied by a chemical plant. Contamination issues identified from site investigation included BTEX compounds, diesel range hydrocarbons, chlorinated solvents and fragrance compounds. Remedial options identified for the odorous soils comprised excavation of gross contamination, containment, ex-situ bioremediation (windrow turning) and thermal desorption. One of the key issues for redevelopment of the site will relate to management of odour during construction and remedial works.

PAINTS, VARNISHES AND INK MANUFACTURE

Contaminants: *Metals and metalloids; Inorganic compounds; Acids/bases; Asbestos; Organic compounds, e.g. aliphatic/aromatic hydrocarbons, halogenated solvents; Fuels; Polychlorinated biphenyls (PCBs); Effluent treatment chemicals.*

This group includes products used both for decorative finishes and protective finishes, i.e. those used in the construction industry and for maintenance, although some paints are used both for decorative and protective purposes. The materials used in the manufacture of coatings are very diverse and include organic and inorganic substances. Pigments and dyes are added to extenders, with binders being used to enable the product to adhere to the surface. Many of the resins and solvents are in liquid form and contamination can occur where these materials are handled or stored.

The production of ink usually involves the use of organic solvents, which can lead to the contamination of groundwater and off-site migration.

Buildings may have been insulated with asbestos as a lagging material or roofing/cladding, some of which may have been disposed of on-site.

Industry Profile: Chemical works, coatings (paints and printing inks) **manufacturing works**

Dealing with volatiles – former paint manufacturing plant

A former paint factory in Lancashire was to be redeveloped for housing use. When the developer first looked at the site it was not possible to undertake a full site investigation as the land was almost entirely covered with buildings. Information provided by the vendors and from a limited investigation showed that paint residues had been deposited in an undeveloped area to the rear of the site, adjacent to a railway line, and had been used to raise the site levels. The limited investigation that the vendor had been prepared to allow through the floors of the buildings confirmed that they were constructed on clay and indicated that paint wastes were confined to one area, where a number of pits had been filled in.

Several developers had rejected the possibility of developing the site as being too risky and costly. Agreement was reached between the vendor and eventual developer based on a price for the land, adjusted to take account of a remediation strategy agreed with the local authority and the Environment Agency. A resident engineer and an environmental surveyor supervised

continued

(*continued*)

the remediation. The final price to be paid was to reflect any saving achieved or additional costs incurred.

The remediation strategy involved removal for off-site disposal of the buried paint residues, mostly in drums many of which had ruptured, and the on-site containment of less contaminated material in an undevelopable part of the site, a tongue of land that had been a rail siding. Clay won from this area was used to re-fill voids on the main part of the site and to create a 'borrow pit' which was lined with an HDPE membrane to create the containment facility. Agreement was reached with the Environment Agency that the work fell within the Exemptions provisions of the Waste Management Licensing Regulations 1994.

PAPER AND PRINTING WORKS, INCLUDING NEWSPRINT (usually excludes 'high street' printers)

Contaminants: *Metals and metallic compounds; Inorganic compounds; Acids and alkalis; Solvents; Other organic compounds incl. organic dyes, ink solvents, cleaning and degreasing solvents, and hydrocarbon fuels; Oils; Asbestos; Polychlorinated biphenyls (PCBs); Inorganic ions.*

This group includes the production of paper and printing. Wood is delivered to the pulp works either as logs or chips, which are then mechanically reduced to a pulp. The pulp then has to be 'washed' and bleached prior to use by the paper making industry. Some paper is also made from waste paper, i.e. newsprint and recycled paper. Such paper has to be de-inked using alkali salts. Bleaching is carried out by the use of hydrogen peroxide, sodium hydrosulphite or formamidine sulphuric acid.

Considerable amounts of water are used in the paper making industry. Paper is made by pressing the pulp onto a mesh screen which bonds the fibres into sheet form and, at the same time, removes the water content. The paper may then be dried, passed between rollers, or coated depending on the quality of product required. Much of the water will be cleaned and re-used but that being disposed to the sewer should be treated beforehand.

A large amount of solid waste is produced, both in the form of sludge from the pulp making process and waste paper. These are usually disposed of to landfill. Contamination can occur from the bleaching products, solvents, coating agents and ink. Some of these can be potentially contaminative to the groundwater. In particular, chlorinated solvents are very persistent and tend to migrate to the bottom of aquifers. Spillages of corrosive materials can have an adverse effect on buildings and their foundations.

Drums and other containers used for the storage of acids etc. may also be potentially contaminative. In the past, some of these containers may have been buried on-site.

Industry Profiles: Pulp and paper manufacturing works.
Profile of miscellaneous industries, incorporating printing and bookbinding works

PESTICIDES MANUFACTURE

Contaminants: *Mineral acids; Alkalis; Asbestos; Organic solvents; Pesticides; Tars; Polychlorinated biphenyls (PCBs); Fuel oils; Effluent treatment chemicals.*

Designed to kill, or render harmless, organisms that may cause damage, pesticides are used in a number of industries, including agriculture, horticulture and forestry, and in the home. They are also used in a wide variety of industrial applications, from rodent control to anti-fouling on ships and pleasure craft. Early pesticides were based on organic salts, mainly arsenates, and sulphur compounds, which were effective as fungicides. The Control of Pesticides Regulations 1986 led to the withdrawal of many pesticides, for health, safety or environmental reasons, but residues of those pesticides, or their ingredients, may be found to have caused contamination on the sites where they were manufactured.

Due to the wide range of applications for pesticides, the nature of ingredients is also extremely diverse, with active ingredients possibly being organic or inorganic, metal or non-metal based, in complex formulations. Solvents or industrial gases may be used as carriers for the active ingredients, or in the production process. Primary processing is, typically, a batch operation in which materials are weighed and mixed or transferred to reaction vessels. These vessels may be heated or cooled as required and the operation is generally computer controlled. Secondary processing involves a number of physical operations, such as milling, granulation, drying, sieving and packaging. Many different products may be produced in the one facility and, at different times, using the same plant and equipment.

Wastes arising from the manufacturing processes may be both liquid and solid. Waste waters are likely to be discharged to the public sewer, probably after undergoing some primary treatment, such as settling out of solids, and pH adjustment. Some more advanced treatments may be employed, such as anaerobic digestion or wet air oxidation, prior to discharge. Hazardous liquid effluents, which cannot be effectively or economically treated on-site prior to discharge, are removed from site to a licensed facility, including landfill or incinerators. Contaminated solid wastes are disposed of to landfill or high temperature incineration, as appropriate.

Sites which have been in use over a long period of time, or which have been used in the past, are more likely to be contaminated than modern plants where stringent environmental controls have been in force throughout the period of operation. Contamination is most likely to be encountered in areas which have been used for the storage of bulk solvents, including underground storage tanks and tanker off-loading areas, or where on-site waste disposal has taken place.

Residues of contaminative substances may be found in underground pipeworks, drainage systems and soakaways. Fuel oil may have been stored on site for space heating and steam-raising, PCB contaminated oils may have leaked from transformers in electricity sub-stations and asbestos may have been used for lagging steam or liquid transfer pipes.

Industry Profile: Chemical works, pesticides manufacturing works

PHARMACEUTICAL INDUSTRIES, INCLUDING COSMETICS AND TOILETRIES

Contaminants: *Metals, metalloids and their components; Inorganic compounds; Acids and alkalis; Asbestos; Organic solvents; Tars from primary processes; Fuel oils and coal; Polychlorinated biphenyls (PCBs); Organic materials inc. surfactants, thickeners, foam stabilisers, pearlescent agents, conditioning agents, fatty alcohols, fatty acids, waxes, cream bases, emulsifiers, humectents, plasticisers, aerosol propellants, anti-microbial agents, sun-screens, dentifrice agents, mineral oils, depilatory agents, fixatives, lubricating oils, perfume synthesis chemicals; Phenols/PAHs.*

The manufacture of pharmaceutical products comprises a primary and secondary stage. Active ingredient processing forms the primary process, with these ingredients then being formed into tablets, ointments etc. for administration, i.e. the secondary stage. Raw materials, such as organic solvents, are generally delivered by tanker for storage on-site. The transfer and storage of such materials can lead to localised contamination due to spillage, leakage etc. The generation of steam may be necessary which can lead to potential contamination from the storage of coal and other fuels.

Where such activities have taken place on a site which overlies an aquifer or where the water table is at such a level that it is at risk of being contaminated from leakage of contaminated materials, there may be cause for concern. Soluble organic materials can migrate leading to a widespread problem.

The production of cosmetics and toiletries are included in this category, which includes skin and hair care products, dental products, nail polish and perfumes. These products are manufactured from both natural and synthetic substances, comprising a wide range of ingredients, some of which may be potentially contaminative. Where such items are delivered in bulk for on-site storage, there is the likelihood of spillage and leaks from tanks or pipes. Again, groundwater can be at considerable risk from the leakage of solvents, which can be highly mobile.

Buildings may have been insulated with asbestos as a lagging material or roofing/cladding, some of which may have been disposed of on-site.

Industry Profiles: Chemical works, pharmaceuticals manufacturing works. Chemical works, cosmetics and toiletries manufacturing works

The importance of a risk-based approach – Pharmaceutical works
A Phase I and Phase II Land Quality Assessment was undertaken of an industrial site near London that for the past 70 years had been used for a mixture of chemical and pharmaceutical manufacture. The Phase I assessment revealed a number of contaminated land uses including tank farms and waste disposal pits in addition to geotechnical characteristics of the site that needed confirming. A major concern was on-site migration of contaminants from an adjacent site that had previously been identified as contaminated by others.

The Phase I study informed the scope and scale of the Phase II intrusive works. The Phase II works demonstrated that on-site migration was not a particular concern. An issue of off-site migration was however identified involving chloroform, a contaminant that was last used on the site some 30 years ago. Using a risk-based approach, and the ground and groundwater conditions actually encountered, it was concluded that the contaminant was at levels that would not significantly harm sensitive receptors. The setting of the site was such that the investigation also identified physical contamination in the form of solution features in an area previously unknown for such natural cavities. Their presence will prevent the use of soakaways and require consideration of appropriate new foundations. The site is currently awaiting disposal.

PLASTIC PRODUCTS MANUFACTURE, MOULDING AND EXTRUSION; BUILDING MATERIALS; FIBREGLASS; FIBREGLASS RESINS AND PRODUCTS

Contaminants: *Metals, metalloids and related compounds; Acids/alkalis; Organic compounds inc. resin/monomer materials, other resins and associated materials, inhibitors/stabilisers, catalysts, plasticisers, promoters/accelerators, sizes/film-forming agents, finishing agents, solvents, oils; Diesel fuel; Biocides; Asbestos; Polychlorinated biphenyls (PCBs).*

Fibreglass products are used for insulation and fire protection products. The production of glass fibre is by the extrusion of molten glass filaments, which are blown by either high-pressure jets of air or steam to produce fine fibres. A binder is sprayed onto the fibres to produce a mat. Waste materials are usually transported to landfills, although waste storage areas may result in localised contamination.

Glass fibre can also be used for resin application when it is then referred to as 'cured' or 'polymerised'. Finished products are chemically stable, although the resins, monomers and stabilisers may also cause localised contamination.

Fuel and lubricating oils are likely sources of contamination in the area of storage tanks.

Industry Profile: Profile of miscellaneous industries, incorporating fibreglass and fibreglass resins manufacturing works

RADIOACTIVE MATERIALS PROCESSING AND DISPOSAL

Contaminants: *Inorganic compounds; Metals; Naturally occurring radioactive materials, e.g. radon and thoron.*

Industries included in this category would be nuclear power stations and recycling works, aircraft manufacturing and maintenance facilities, watch and instrument manufacture, hospital waste including X-ray equipment and use and radiation for treatment purposes, laboratory waste and mining wastes (uranium). However, medical facilities are not included within this guide, nor are nuclear power stations due to the high degree of specialist knowledge that would be required in dealing with such contamination. It is unlikely that radioactive materials would be encountered on other industrial sites apart possibly from the use of luminous paint. Luminous paint was used on aircraft instruments during World War II and also on minefield markers.

In order to lessen the risk from exposure to radiation, the International Commission on Radiological Protection made three key recommendations:

(1) Individuals should not be exposed to radiation unless sufficient benefits are produced to justify this.
(2) Radiation levels should be kept as low as possible.
(3) Exposure limits should be set above which the risk would be unacceptable.

Low-level radioactive waste may have been disposed of to pre-licensed landfill sites and caution must be exercised when undertaking site investigations or excavation work on such sites due to the risk of carcinogenic and genetic damage. Discharges from major nuclear sites also require authorisation from MAFF or the Secretaries of State in Wales and Scotland.

In some areas of the country, particularly where granite occurs, radiation may be present in the ground as a gas known as radon. This is a colourless, odourless gas, which can be dangerous where it becomes trapped in buildings, and a high concentration builds up.

Advice and guidance can be sought from the Environment Agency and the National Radiological Protection Board (NRPB). Also see the Radioactive Substances Act 1993.

Industry Profile: None

RAILWAY LAND, INCLUDING YARDS AND TRACKS

Contaminants: *Fuel oils; Lubricating oil; Paraffin; Polychlorinated biphenyls (PCBs); Polyclyclic aromatic hydrocarbons (PAHs); Solvents; Ethylene glycol; Creosote; Herbicides; Ferrous residues; Metal fines; Asbestos; Ash; Sulphate.*

Contamination can occur through the construction processes employed in the establishment of the railway tracks, through the operation of engineering depots, goods yards and sub-stations.

Building railway tracks can involve the creation of cuttings and embankments, and the construction of bridges and tunnels. In some cases, natural material derived from the creation of cuttings would be used to form the embankments but in addition, large quantities of clinker ash from local industries were used. Crushed slag is frequently used as ballast to support the track, as is steam locomotive ash.

The electrification of the rail system involved transforming electricity from the National Grid to a lower voltage by means of sub-stations. These sub-stations formerly used oils containing polychlorinated biphenyls (PCBs) although these have now been replaced.

Engineering depots were used for a variety of tasks associated with the maintenance and repair of the tracks which could result in soil contamination by fuel oil, lubricating oil, solvents, herbicides, paint, anti-freeze, etc.

Goods yards are frequently situated adjacent to railway stations and were mostly used for the storage of coal, although a variety of other products have been, and are currently, stored in goods yards. Cattle and other domestic animal loading and transfer operations may have taken place in some yards. 'Break of bulk' operations may also have taken place.

Buildings may have been insulated with asbestos as a lagging material or roofing/cladding, some of which may have been disposed of on-site.

Industry Profile: Railway land

Correct identification of contaminant sources – Rail sidings and oil depot
A site within a major conurbation had rail sidings used to stage and refuel diesel multiple units (DMUs) until the line was electrified. Facilities still remained to fuel diesel engines. Adjacent commercial land (hydrogeologically upstream of the commercial sites) which included an oil depot, electroplating and car servicing land, all became zoned for housing as industries closed.

continued

(*continued*)

Investigating the oil depot and the car service area for housing revealed widespread organic contamination on the perched groundwater, and some in the underlying chalk aquifer. By contacting the local authority, the consultants to the housing developer became aware that there was a past history of diesel spills on the railway land which had been the subject of council 'action' in a previous decade, and that an interceptor trench was constructed within the rail yard boundary to collect spillages.

Having first assumed that the organic contamination under the oil depot was 'self-sourced', further testing and 'fingerprinting' revealed widespread PCB contamination as well as TPH, which historically could be related to fuels used on the DMUs and not that stored for domestic sale on the oil depot. A significant contribution to the land remediation costs was made to the developer by the adjacent landowner/polluter. Remedial work continued to collect and treat the railway site, and a bentonite barrier wall 'protects' the commercial sites from further infiltration.

RUBBER PRODUCTS AND PROCESSING

Contaminants: *Metal and metalloid contaminants inc. activators, desiccants, colouring pigments; Inorganic compounds inc. flame retardants; Organic Compounds inc. rubber precursors, rubber, rubber preservatives, vulcanising agents, processing aids, plasticisers, and extenders, accelerators, activators, stain protectors, retarders, waxes, tackifying resins, hardeners, peptisers; Others including vulcanising agents, fillers, anti-degradants.*

Both natural and synthetic rubbers are used in the manufacture of such items as road vehicle tyres, aircraft and industrial tyres, footwear and sundry industrial components. The production of latex products is also included in this category. The raw materials are delivered in the form of rubber in bales, whilst latex is delivered by tanker. Other raw materials include preservatives, fillers in the form of carbon black, accelerators, activators such as zinc oxide, solvents containing toluene, xylene, benzene etc. Rubber itself is a highly combustible material and flame-retardants are used containing zinc metaborate dihydrate, sodium borate and sodium metaborate. Carbon black is not biodegradable and may be associated with polyclyclic aromatic hydrocarbons (PAHs).

The manufacture of vehicle tyres accounts for the majority of the use of rubber. Prior to the manufacturing process, the raw ingredients are mixed by combining the rubber and compounding ingredients. Cotton, nylon or steel may then be added, followed by a moulding process. Rubber can also be extruded, as for tyre treads.

Contamination is most likely to occur from the additives used, solvents and fuel oils. The manufacturing process uses steam, which may require fuel oil or coal to be stored on site. Electrical sub-stations may be necessary which could have used polychlorinated biphenyls (PCBs). Buildings may have been insulated with asbestos as a lagging material or roofing/cladding, some of which may have been disposed of on-site.

Industry Profile: Chemical works

- rubber processing works (including works manufacturing tyres or other rubber products)

SCRAPYARDS

Contaminants: *Metals and metalloids; Inorganic compounds; Acids inc. hydrochloric, phosphoric, sulphuric (from vehicle batteries); Alkalis; Asbestos; Organic compounds inc. fuels, hydraulic oils, lubricating oils, paints, phenols, polychlorinated biphenyls, solvents; Others inc. radioactive materials, biodegradable items.*

Contamination can be widespread on such sites as a result of the number of different activities that take place in the dismantling, storage and recovery of scrap materials. Each scrap site is required, by law, to hold a record of the types of metals handled.

Contamination can occur from the breaking up of vehicles where oils and fuel from scrap vehicles are not first removed. Non-metal parts should be properly disposed of but are sometimes burnt on-site instead, resulting in contaminated ash. Asbestos can be released from vehicle brake pads.

Battery dismantling is usually carried out by specialist firms or they may be recycled, although in the past batteries stored on site may result in leakages of acid into the soil. Radioactive materials can result from the disposal of such items as aircraft instrument panels incorporating radioactive paints.

Industry Profile: Waste recycling, treatment and disposal sites

- metal recycling sites

On-site disposal of waste oils – vehicle dismantlers

A riverside scrapyard location was examined for conversion into a public amenity area. Relevant receptors were the adjacent tidal river, recently recovering with returned fish species, and the underlying chalk aquifer. Although 75% of the river boundary had functioning retaining walls, there was a large area of end-tipped rubbish in the riverbank. Initial trial pits revealed two soft areas in this zone, one of which the leg of the mechanical excavator sank into; these turned out to be intact buried cars!

The site also had two river outfalls, apparently connected to the highway drainage system, of which only one was visibly functioning at rainfall events. Enquiry of the council depot revealed a knowledgeable employee, who stated that the second outfall silted up and a soakaway had been constructed on the scrapyard in the 1960s for discharge to the river Gravels. Tracing back of the outfall from the river by excavator revealed the soakaway, which was surrounded by oily gravels and odorous waters. Uncovering the top of the soakaway revealed the disposal point of all the engine oil discards from the scrapping works for the past twenty years.

SEWAGE TREATMENT WORKS

Contaminants: *Metals, metalloids and their compounds; Inorganic compounds; Acids/alkalis; Asbestos; Organic compounds; Polychlorinated biphenyls (PCBs) and other transformer oils; Micro-organisms (pathogens); Methane; Carbon dioxide; Hydrogen sulphide.*

Modern sewage treatment works use the activated sludge technique, which results in the solid matter being separated from the water content and treatment with microorganisms. Foul sewage, containing only domestic and industrial effluent, is delivered to the treatment works. Following the separation process, which involves primary and secondary treatment, the solid waste or sludge is used on farms, disposed of to landfill or incinerated.

Care must be exercised in the disposal of sludges that may contain pathogenic microorganisms such as *Salmonellae* and *Typhus* bacteria. These sludges are the most common cause of contamination. Other wastes such as grit, stones and solid matter removed in the preliminary screenings are usually disposed of to landfill.

However, other sources of contamination are the pipes, tanks and treatment beds and any leakages. Sludges can result in methane and carbon dioxide production and dried sludge can be combustible.

Buildings may have been insulated with asbestos as a lagging material or roofing/cladding, some of which may have been disposed of on-site.

Industry Profile: Sewage works and sewage farms

Logistics and Health and Safety – Sewage works site investigation
Redevelopment of a 19th century sewage works required preliminary investigation in, around and through the existing structures, including sewage sludge lagoons. Historical plans of the multi-level structures were limited. Residual sewage sludge in some parts restricted access. Investigations required innovative uses of typical investigation techniques combined with a flexible approach to react to unforeseen findings.

The limited available data were combined with structural assessments to identify suitable techniques for investigation. This used a combination of techniques such as drilling using both rotary and shell and auger methods through the structure and in sludge lagoons, window sampling through piers and trial pitting using a long-reach excavator. Logistical issues requiring consideration included the weight of plant and the investigation position relative to likely structures as well as the usual constraints such as the

continued

(continued)

location of abundant services and the presence of overhead pylons. Rigorous methodologies and health and safety plans were also formulated to clearly define all stages of the works.

Information gathered was used for a Phase II Land Quality Assessment which indicated that ground and groundwater contamination was limited and remedial action was not required. Incidental information obtained about the structure was also used in the detailed design for the demolition of the structure, which included removal of the sewage sludge. All information was used as part of an Environmental Impact Assessment submitted as part of the planning application for the redevelopment of the site.

TANNING AND LEATHERWORKS

Contaminants: *Metals and metalloid compounds inc. dyeing and tanning agents, chromium-containing tanning agents, potassium alum (an obsolete tanning agent that may be found on older sites); Inorganic compounds inc. sulphides, others, biocides; Acids and alkalis; Organic compounds inc. vegetable oils, oligomers of formaldehyde, phenolic materials or naphthalene, oil tans, insecticides, solvents, lubricants, dyestuffs, binders, enzymes and lacquers, fuel oils.*

The tanning process, which involves cleaning and treating the animal skin, used to be carried out by hanging the hide in deep pits. This process has now been replaced by a mechanical agitation system. Tanning agents include chromium (trivalent), sulphate, vegetable and synthetic agents.

Wastes mostly comprise liquid effluent discharged to sewer following treatment. The solid waste is usually disposed of to landfill. Various insecticides can also be present in the waste, such as permethrin and cypermethrin, both of which are used in sheep dips. Pyrethrum and boric acid may be used on both sheep and cattle and therefore present in the effluent produced as a result of the cleaning of skins and hide.

As many skins and hides are imported it is not possible to state which biocides may have been used. Lindane, whilst not used in the United Kingdom as an insecticide, is still used in South America and is frequently present on imported items. However, lindane will usually disappear from soil within 3–10 years.

Other sources of contamination are leakages and spills of chemicals being stored and metallic wastes from the tanning process.

Industry Profile: Animal and animal products processing works

The blighting effect – tannery

A landowner had his property blighted as a result of part of its former land use. The Town Plan of 1851 recorded the north-eastern corner of the site as a tannery. Some 32 tanks were recorded on this plan along with the Tan and Engine House. The site is recorded as a tannery up to the OS Map edition of 1954. Derelict for a number of years the owner had a local house builder interested but because of the uncertainty over potential contamination issues, including the perception of anthrax in the public mind, a poor offer had been received.

continued

(continued)

Appointed by the site owner, a consultant undertook a Phase I and II Land Quality Assessment to confirm the geotechnical and geoenvironmental status of the site. The investigation revealed that for a residential end use, the majority of the site (85%) had no abnormal development costs whilst limited remediation was required on the remaining part of the site. Anthrax was not proven and the site was subsequently sold for an additional £1.7 million.

TAR, BITUMEN, LINOLEUM, VINYL AND ASPHALT WORKS

Contaminants: *Metals and metalloids; Inorganic compounds; Organic compounds; Solvents; Fuels; Asbestos; Polychlorinated biphenyls (PCBs) Polyclyclic aromatic hydrocarbons (PAHs).*

In the manufacture of linoleum, vinyl and bitumen-based floor coverings, contamination can occur through both the production facilities and through spillages and disposal of waste materials. Waste water contaminated with solvents is disposed of by tanker but may in the past have been disposed of to soakaways.

Potential contaminants can include toluene, white spirit and other solvents, pigments and stabilisers.

Significant amounts of power, notably steam, are required which may be produced by fuel oil, coal, natural and petroleum gas. On-site electricity substations containing polychlorinated biphenyls (PCBs) may also be present.

Buildings may have been insulated with asbestos as a lagging material or roofing/cladding, some of which may have been disposed of on-site.

Industry Profiles: **Ceramics, cement and asphalt manufacturing works**
Chemical works – linoleum, vinyl and bitumen–based floor covering manufacturing works

TEXTILES MANUFACTURE AND DYEING

Contaminants: *Metals, metalloids and their compounds; Inorganic compounds; Sizing agents; Organic solvents; Other organic compounds; Polychlorinated biphenyls (PCBs); Dyes; Pesticides; Asbestos; Anthrax spores; Carbon dioxide; Methane.*

Contamination can occur from the delivery and storage of chemicals used in textile manufacture. A further source of contamination can arise from the washing of yarns and fabric, which may release effluents into the foul sewer. Such effluent may be leaked onto the site or may have been discharged to local waterways, and may contain pesticides such as lindane and pentachlorophenol. If bleaching has taken place, the effluent may contain such contaminants as sodium hypochlorite and hydrogen peroxide.

The processes used in textile manufacture and dyeing require considerable consumption of power. Coal fired boilers have now mostly been replaced by gas or oil. Coal fired boilers would produce residues of boiler ash containing heavy metals and sulphates, whilst oil would require on-site fuel storage tanks. Electricity generators may have used polychlorinated biphenyls (PCBs).

Older properties used for textile manufacture, textile works and dye works may contain asbestos as lagging material or roofing/cladding, some of which may have been disposed of on-site.

Industry Profile: Textile works and dye works

TIMBER TREATMENT WORKS

Contaminants: *Metals and metalloids; Inorganic compounds; Acids/alkalis; Asbestos; Organic solvents, e.g. white spirit, kerosene; Preservatives; Polychlorinated biphenyls (PCBs); Dioxins and furans; Fuels.*

Contamination can occur from the use of preservatives such as creosote and salt-based materials in the treatment of sawn timber. Copper–chromium–arsenic (CCA) solutions were introduced in the 1930s and their use continues to this day, together with light organic solvents solutions. Timber treatment facilities are frequently close to sawmills and timber products manufacturing works. However, the industry has declined in recent years.

The ingredients for the preparation of timber treating solutions can comprise liquid or pastes and are usually stored on site in either tanks or drums. For instance, creosote is usually delivered as a liquid, by tanker. Creosote is a derivative from coal tar and is only suitable for external use due to the strong odour. Such uses would include fencing, telegraph poles, etc. Diluted forms were available for domestic use.

CCA preservatives are usually delivered as a water based solution or paste, for application by industrial high pressure/vacuum systems. As CCA is odour free, it can also be employed to treat timber used in the construction of buildings, i.e. roof timbers, joists etc.

Light organic solvents comprise a number of active ingredients dissolved in white spirit, kerosene or other petroleum distillates. Pigments can be added to the solvents which are suitable for use on machined components.

Contamination within timber treatment works or sawmills is likely to be associated with contaminated sawdust or in the storage/transfer areas for the preservatives. Older sites may also have underground storage tanks, soil soak-aways and waste disposal areas, all of which may lead to contamination.

Other potential sources of contamination may include the use of fuel oil, petrol, or diesel and on-site electricity generation.

Industry Profiles: Timber treatment works
Timber products manufacturing works
Chemical works: pesticides manufacturing works

Identifying different phases of development – long-established timber treatment yard

In a tidal river location was a historic timber treatment yard that had functioned since Napoleonic times but at the time of the investigation was only selling garden sheds and greenhouses. Investigation of the site revealed a strange jumble of intermittent organic contamination at several levels within the four to six metres of made ground overlying the river gravels and chalk.

The pattern of such contamination (not speciated in any great detail in relation to the history) was only revealed when the site excavation took place. A variety of infilling materials and three previous wharfages (two with anchorages) were exposed. Discard chalk waste tended to absorb the contaminants, and change colour as an indication of 'strength', whilst 'pools' of floating liquors were constrained behind successive timber (two types), concrete and steel sheetpile walls, at differing levels and in the granular backfill around the anchor ties (some of the ties were so well corroded, it was a surprise the wall still stood).

The site was successfully redeveloped with decontamination costing £750 000 of the £2 750 000 works, and included a new river wall with inclined anchorages (drilled before decontamination, so that the earlier river walls could be removed safely). The redevelopment won a regional ICE award, and garden sheds are still sold on an adjacent 'inland' site; their former location now being a floodlit five-a-side pitch and part of the regional coastal pathway.

TRANSPORT DEPOTS, ROAD HAULAGE, COMMERCIAL VEHICLE FUELLING, LOCAL AUTHORITY YARDS AND DEPOTS

Contaminants: *Metals and their compounds including lead, chromium, zinc, copper, vanadium; Acids/alkalis; Asbestos; Organic compounds including non-halogenated solvents, halogenated solvents; Phenols/polycyclic aromatic hydrocarbons (PAHs).*

Road haulage and distribution centres are situated throughout the country and range in size from parking provision for a single vehicle owner operator to large-scale distribution centres with hundreds of vehicle movements every day. Many sites may include fuelling and vehicle maintenance facilities, although many larger operators have tended towards greater centralisation, or outsourcing, of these functions in recent years. Nevertheless, historic contamination may remain on sites where these functions have been discontinued. Vehicle wash facilities are another potential source of contamination, unless waste water is disposed of via an effective petrol interceptor.

Raw materials for use in connection with vehicle operations include diesel fuel, lubricating oils, cleaning materials, brake fluids and anti-freeze. Repair and maintenance facilities will generate wastes associated with operations such as welding, soldering and re-spraying, which may include body filler materials (e.g. glass fibre) solvents and paints. On some older sites' waste oils, which can contain heavy metals such as lead and vanadium, may have been disposed of by simply pouring them on the ground, or into underground structures, such as redundant air-raid shelters. Modern disposal methods usually involve direct piping of waste oils from servicing facilities to holding tanks to await collection by specialist contractors.

Used tyres and other combustible wastes may have been burned on-site, leaving behind PAH contamination. Today specialist contractors generally undertake tyre repair or replacement, with used tyres being returned to the contractor or disposed of to landfill. Combustible materials, such as used packaging, is generally transferred to a baling plant, or consigned to a skip, for recycling.

Frozen food storage and distribution centres have large cold and chill stores that use ammonia as a refrigerant. Refrigerated trailers use liquid nitrogen as a refrigerant. A wide range of liquid and powder materials is transported by road tankers, ranging from food products to highly toxic or corrosive chemicals. Tank cleaning operations may result in ground contamination.

General warehousing operations, where no vehicle servicing, fuelling or tank washing operations are located on sites can, nevertheless, produce contamination.

Accidental occurrences, such as split drums or sacks, may result in unknown materials finding their way into the ground. Such incidents may not be recorded and may only come to light during the course of a site investigation.

Industry Profile: Road vehicle fuelling, service and repair

- transport and haulage centres

Imported materials to create hardstanding for vehicles
A site formerly occupied by a road haulage firm was to be redeveloped for housing. No other use had been recorded on the site. The site investigation revealed extensive areas of made ground, containing ashes and tarry wastes, subsequently identified as being gasworks residues. Extremely high concentrations were also found of petroleum hydrocarbons, toluene extractable matter, zinc, cyanide and sulphate. High levels of organic contamination were found in one area.

The possibility of retaining at least some of the contamination on site, under a large landscaped area, was considered but was rejected due to the nature of the contamination. The contaminated material was therefore removed under stringent health and safety supervision and was disposed of as special waste (see Syms & Knight, 2000, Chapter 11 for more details).

WASTE DISPOSAL SITES, INCLUDING HAZARDOUS WASTES, LANDFILLS, INCINERATORS, SANITARY DEPOTS, DRUM AND TANK CLEANING, SOLVENT RECOVERY

Contaminants: *Can include non-toxic inorganic chemicals; Solvents and oils; Resins, paints and organic sludges; Organic chemicals; Chlorinated hydrocarbons; Polychlorinated biphenyls (PCBs); Phenols; Landfill gas and leachates; Metals and their compounds inc. aluminium sulphate, iron sulphate, sodium nitrate; Acids inc. sulphuric, hydrochloric; Alkalis; Asbestos; Organic compounds inc. diesel, paraffin, solvents, interceptor waste, polychlorinated biphenyls (PCBs), detergents; Oils.*

The need for effective disposal of wastes has been recognised since at least the mid-19th century, being regulated by Acts of Parliament and local authority byelaws. Disposal and treatment systems include landfill sites, incinerators, the first successful one of which was constructed in Manchester in 1876, and waste transfer stations, which have operated since around the beginning of the 20th century.

A wide variety of sites have been used as landfills, including low-lying land and marshland, former quarries, disused canals and railway cuttings, poor agricultural land and 'flashes' created by mining subsidence. Until only relatively recently, little consideration was given to the structural integrity of landfill sites, as a result of which clay or synthetic liners may be absent from older sites. Site licensing was only introduced in 1976 and therefore records relating to older sites may be difficult, or even impossible, to obtain. Leachate and landfill gas, arising out of the decomposition of waste materials, are the major problems associated with landfill sites and the process of decomposition may take decades.

The main function of early municipally owned incinerators was to reduce the bulk of waste, with the ashes then being used for road construction or other building purposes. Unsuitable incineration residues were sent to landfill. Little consideration was given to efficiency of the combustion process or to heat recovery, both of which (together with emission control) are objectives in the design of modern incinerators.

Waste transfer stations collect and process a significant proportion of domestic and commercial wastes, together with some industrial wastes. As with landfills and incinerators the primary transportation method is by road but some wastes are transported by rail or barge. Modern collection points (civic amenity sites) for the disposal and recycling of wastes also fall within the description of waste transfer stations.

Hazardous waste treatment plants process and recycle a wide variety of chemicals and other liquid wastes using a number of different treatment techniques. These include incineration and pyrolysis, neutralisation, precipitation, oxidation, filtration and stabilisation. Biological treatment may be used to break down degradable organic based wastes. Liquid waste materials are delivered to the treatment plant in tankers or drums and solid wastes in bulk containers or large bags made of woven fibre with a plastic membrane. Filter cake from treatment operations may be transported by skip.

Approximately five million steel drums are recycled every year in the United Kingdom, of which 800 000 are cleaned by high temperature heating and the rest are washed. The drums are normally empty when received for recycling but residues are inevitably brought onto the site. Plastic drums are also recycled, around 300 000 each year, and extreme care needs to be taken in their re-use as plastic can absorb chemicals.

Metal wastes received at recycling sites may be contaminated with oils and greases, requiring degreasing with solvents such as trichlorethylene or methyl ethyl ketone. Historically, before the regulation of trade effluent came into force, the on-site disposal of spent solvents may have resulted in ground contamination. Water is also used to wash material and may be recycled or discharged to drains. The dismantling of batteries is now carried out at specialised recycling plants, but some older plants may be seriously affected by lead contamination of the soil.

A very wide range of materials, in both solid and liquid forms, may have been treated or disposed of at waste recycling plants. In many cases, especially closed sites or plants which have been in use over a long period of time, it may not be possible to obtain much information as to the types of materials which have been treated or buried on the site. The types of industry (both current and historic) in the neighbourhood of a treatment facility may give an indication of the wastes that have been handled but some facilities take waste from a wide area.

Activities and treatment processes on waste handling sites may have changed over time, as too may have the physical location of treatment plant and buildings. Older facilities may have lacked adequate containment areas for liquids and other wastes prior to treatment and, although these may have been substantially improved, residual problems may remain.

Industry Profiles: Waste recycling, treatment and disposal sites

- hazardous waste treatment plants
- landfills and other waste
- treatment or waste disposal sites
- drum and tank cleaning and recycling plants
- solvent recovery works

References

Albrechts, L. (2002) The Planning Community Reflects on Enhancing Public Involvement. Views from Academics and Reflective Practitioners. *Planning Theory and Practice*, vol. 3, no. 3, pp. 331–347.

Arup Economics and Planning with the Bailey Consultancy (2002) *Resourcing of Local Planning Authorities*. DTLR, London.

Ashworth, W. (1954) *The Genesis of Modern British Town Planning*. Routledge and Kegan Paul, London.

Atkinson, D. (ed.) (1995) *Cities of Pride: Rebuilding Community, Refocusing Government*. Cassell, London.

Bell, R. (1998) The Impact of Detrimental Conditions on Property Values. *The Appraisal Journal*, vol. LXVI, no. 4, pp. 380–391.

Briggs, A. (1983) *A Social History of England*. Weidenfeld and Nicolson, London.

Brooks, M. P. (2002) *Planning Theory for Practitioners*. APA, Chicago, Illinois.

Campanella, J. (1984) Valuing Partial Losses in Contamination Cases. *The Appraisal Journal*, April, pp. 301–304.

Castells, C. (1989) *The Informational City*. Blackwell, Oxford.

Cherry, G. (1988) *Cities and Plans*. Edward Arnold, London.

CIRIA (1995–1998) *Remedial Treatment for Contaminated Land*, in 12 volumes. Construction Industry Research and Information Association, London.

Coughlin, D. J. (1995) Real Estate Development. *The Handbook of Real Estate Portfolio Management*. Irwin Professional Publishing, Burr Ridge, Illinois, pp. 309–339.

Council for British Archaeology (2003) Letter containing comments in response to the ODPM's consultation in respect of proposed changes to Part C of the Building Regulations, CBA, York, March.

CPRE (Campaign to Protect Rural England) (2003) Time for the Government to Raise its Sights on Brownfield Building. News release 29 May.

DEFRA (Department for Environment, Food and Rural Affairs) (2002) Letter dated 20 December headed 'Withdrawal of ICRCL Guidance Note 59/83' (2nd edn), (ref CLAN 1/02) and accompanying briefing note (ref CLAN 3/02).

DEFRA (Department for Environment, Food and Rural Affairs) (2003a) *Interim Report of the Waste Permitting Review*. Waste Management Division, Branch 6, DEFRA, London.

DEFRA (Department for Environment, Food and Rural Affairs) (2003b) *Proposal for an EU Directive on Environmental Liability*. Consultation document, Europe Environment Division, DEFRA, London.

DEFRA (Department for Environment, Food and Rural Affairs) (2003c) *Proposal for an EU Directive on Environmental Liability*. Consultation covering letter, Europe Environment Division, DEFRA, London.

DETR (Department of the Environment, Transport and the Regions) (1997) *Building Partnerships for Prosperity: sustainable growth, competitiveness and employment in the English regions*. Cm 3814. The Stationery Office, London.

DETR (Department of the Environment, Transport and the Regions) (1998) *Modernising Planning*. DETR, London.

DETR (Department of the Environment, Transport and the Regions) (2000a) *Planning Policy Guidance Note 3: Housing*. The Stationery Office, London.

DETR (Department of the Environment, Transport and the Regions) (2000b) *Environmental Protection Act 1990: Part IIA Contaminated Land*. Circular 02/2000. DETR, London.

DETR (Department of the Environment, Transport and the Regions) (2001) *PPG13: Transport*. DETR, London.

DoE (Department of the Environment) (1994) *Planning Policy Guidance Note 23 – Planning and Pollution Control*. HMSO, London.

DoE (Department of the Environment) (1994–1995) *Contaminated Land Reports 1–6*. HMSO, London.

DoE (Department of the Environment) (1996) *Household Growth: Where Shall We Live?* Cm 3471. HMSO, London.

DoE (Department of the Environment) (1997) *Planning Policy Guidance 1: General Policy and Principles*. DETR, London.

DTI (Department of Trade and Industry) (2002) *Modernising Company Law*, a White Paper, Cm 553-1. www.dti.gov.uk/companiesbill/prelims.pdf

DTLR (Department of Transport, Local Government and the Regions) (2001) *Planning: Delivering a Fundamental Change*. DTLR, London.

DTLR (Department of Transport, Local Government and the Regions) (2002) *Development on Land Affected by Contamination*. Consultation paper on draft Technical Guidance Note. DTLR, London.

English Partnerships (2003) *Towards a National Brownfield Strategy*. Research findings for the Deputy Prime Minister. English Partnerships, the National Regeneration Agency, September.

Environment Agency (2002a) *Assessment of Risks to Human Health from Land Contamination: an overview of the development of Soil Guideline Values and related research*. Contaminated Land Report 7. Environment Agency, Bristol.

Environment Agency (2002b) *Priority Contaminants for the Assessment of Contaminated Land*. Contaminated Land Report 8. Environment Agency, Bristol.

Environment Agency (2002c) *Contaminants in soil: collation of toxicological data and intake values for humans*. Contaminated Land Report 9. Environment Agency, Bristol.

Environment Agency (2002d) *The Contaminated Land Exposure Assessment (CLEA) Model: technical basis and algorithms*. Contaminated Land Report 10. Environment Agency, Bristol.

Environment Agency (2003a) *Land Quality: an Introduction to CLEA*. www.environment-agency.gov.uk/subjects/landquality

Environment Agency (2003b) *Model Procedures for the Management of Land Contamination*. Consultation draft 2V2. Contaminated Land report 11 (CLR11). Environment Agency, Bristol.

Environmental Protection Act 1990, Part IIA, the 'contaminated land' legislation, introduced via s.57 of the *Environment Act 1995*. The Stationery Office, London.

Financial Services Authority (2003) *Combined Code on Corporate Governance*. FSA, London. www.fsa.gov.uk/pubs/ukla/lr_comcode2003.pdf

Graham, S. & Marvin, S. (1996) *Telecommunications and the City*. Routledge, London.

Harley, J. B. (1975) *Ordnance Survey maps, a descriptive manual*. HMSO, London.

Hayter, R. (1997) *The Dynamics of Industrial Location*. John Wiley and Sons, Chichester.

HM Treasury (2001) Budget 2001, chapter 6, *Protecting the Environment*. HM Treasury, London.

HM Treasury (2003) Budget 2003, chapter 7, *Protecting the Environment*. HM Treasury, London.

House of Commons (1990) *Contaminated Land*. First Report of the Select Committee on the Environment. HMSO, London.

ICRCL (1987) *Guidance on the assessment and redevelopment of contaminated land*. ICRCL 59/83, second edition 1987. Interdepartmental Committee on the Redevelopment of Contaminated Land, London.

International Valuation Standards Committee (2000) *International Valuation Standards*, London.

Jackson, T. (2000) The Effect of Previous Environmental Contamination on Industrial Real Estate Prices. Paper presented at the Valuation 2000 conference, Las Vegas, July, pp. 59–68.

Keeble, L. (1964) *Principles and Practice of Town and Country Planning*. Estates Gazette, London.

Kennedy, P. J. (1998) *Investment Valuation of Contaminated Land and UK Practice: A Study with Special Reference to Former Gasworks*. PhD thesis, The Nottingham Trent University.

Kitchen, T. (1997) *People, Politics, Policies and Plans*. Paul Chapman, London.

Kitchen, T. (2001) Planning in Response to Terrorism: the Case of Manchester, England. *Journal of Architectural and Planning Research*, vol. 18, no. 4, pp. 325–340.

Kitchen, T. (2002) The Balance between Certainty, Speed, Public Involvement and the Achievement of Sustainable Development in the Planning System:

the Impact of the Planning Green Paper. In: Kitchen, T. (ed.) *Certainty, Quality, Consistency and the Planning Green Paper: Can Planning Deliver the Goods?* Yorkshire Conference Series Partners, Sheffield, pp. 21–34.

LaGrega, M. D., Buckingham, P. L., Evans, J. C. & The Environmental Resources Management Group (1994) *Hazardous Waste Management*. McGraw-Hill, New York.

Lizieri, C., Palmer, S., Finlay, L., & Charlton, M. (1995) Valuation Methodology and Environmental Legislation: a research project for the RICS Education Trust, City University Business School, discussion paper series. City University, London.

Manchester City Council (1995) *Manchester: 50 Years of Change*. HMSO, London.

Martin, I. & Bardos, P. (1996) *A Review of Full Scale Treatment Technologies for the Remediation of Contaminated Soil*. Report for The Royal Commission on Environmental Pollution. EPP Publications, Richmond.

McCarthy, P., Prism Research & Harrison, T. (1995) *Attitudes to Town and Country Planning*. HMSO, London.

McNulty, T. (2003) *Putting Planning First, Culture Change for the Planning Profession*. ODPM, London.

Mitchell, P. S. (2000) Estimating Economic Damages to Real Property Due to Loss of Marketability, Rentability and Stigma. *The Appraisal Journal*, vol. LXVIII, no. 2, pp. 162–170.

Mundy, B. (1992a) Stigma and Value. *The Appraisal Journal*, January, pp. 7–13.

Mundy, B. (1992b) The Impact of Hazardous Materials on Property Value. *Appraisal Journal*, April, pp. 155–162.

NLUD (National Land Use Database) (2003) *Previously Developed Land that may be available for redevelopment (Brownfield Sites) in 2002*. Report, September. www.nlud.org.uk

ODPM (Office of the Deputy Prime Minister) (2002a) *Your Regions, Your Choice*. Cm 5511. HMSO, London.

ODPM (Office of the Deputy Prime Minister) (2002b) *Sustainable Communities: Delivering Through Planning*. ODPM, London.

ODPM (Office of the Deputy Prime Minister) (2002c) *Land Use Change in England: LUCS17*. ODPM, London.

ODPM (Office of the Deputy Prime Minister) (2003a) *Sustainable Communities: Building for the Future*. ODPM, London.

ODPM (Office of the Deputy Prime Minister) (2003b) *Land Use Change in England: LUCS18*. ODPM, London.

ODPM (Office of the Deputy Prime Minister) (2003c) *Decontaminated Land to be brought back into use thanks to State Aid Approval*. ODPM News Release 2003/0157: 30 July.

ODPM (Office of the Deputy Prime Minister) (2003d) *Possible Future Performance Standards for Part L*. Response following formal consultation. ODPM, London, October.

ODPM (Office of the Deputy Prime Minister) (2004) *Draft Regional Assemblies Bill: Policy Statement*. ODPM, London.

Patchin, P. (1988) Valuation of Contaminated Properties. *Appraisal Journal*, January, pp. 7–16.

Patchin, P. (1992) *Value Estimation for Impaired Properties*. Proceedings of the Appraisal Institute Symposium, pp. 139–140.

Patchin, P. (1994) Contaminated Properties and the Sales Comparison Approach. *Appraisal Journal*, vol. 62, no. 3, pp. 402–409.

Petts, J., Cairney, T. & Smith, M. (1997) *Risk-Based Contaminated Land Investigation and Assessment*. John Wiley & Sons, Chichester.

Planning Advisory Group (1965) *The Future of Development Plans*. HMSO, London.

Planning Officers Society (2002) *A Guide to Best Value and Planning*. Planning Officers Society, no publication location specified.

Planning Officers Society (2003) *Moving Towards Excellence in Planning*. Planning Officers Society, no publication location specified.

Richards, T. O. (1995) *A Changing Landscape: The Valuation of Contaminated Land and Property*. College of Estate Management Research Report, Reading.

Richards, T. O. (1996) The Valuation and Appraisal of Contaminated Land and Property. Paper presented at The Cutting Edge 1996: a research conference of the Royal Institution of Chartered Surveyors, September. The University of the West of England, Bristol.

Richards, T. O. (1997) *Is it Worth the Risk? The Impact of Environmental Risk on Property Investment Valuation*. College of Estate Management Research Report, Reading.

RICS (Royal Institution of Chartered Surveyors) (1995) Valuation Guidance Note 2, in the *RICS Appraisal and Valuation Manual*. RICS, London.

RICS (Royal Institution of Chartered Surveyors) (2003a) *RICS Appraisal and Valuation Standards*. Available online or in loose leaf format, RICS Books, Coventry.

RICS (Royal Institution of Chartered Surveyors) (2003b) *Contamination and Environmental Matters: their implications for property professionals*. RICS guidance note published for the Environment Faculty, RICS Books, Coventry.

Savitch, H. V. (1988) *Post-Industrial Cities*. Princeton University Press, Princeton, New Jersey.

Simons, R. (1998) *Turning Brownfields into Greenbacks*. pp. 138–139.

Skeffington, A. (Committee on Public Participation in Planning, chaired by Arthur Skeffington MP) (1969) *People and Planning*. HMSO, London.

Syms, P. (1997a) *Contaminated Land: the practice and economics of redevelopment*. Blackwell Science, Oxford.

Syms, P. (1997b) *The Redevelopment of Contaminated Land for Housing Use*. Research report supported by the Joseph Rowntree Foundation, ISVA, London.

Syms, P. (1999) *Desk Reference Guide to Potentially Contaminative Land Uses*. ISVA, London.

Syms, P. (2001) *Releasing Brownfields*. Research report prepared for the Joseph Rowntree Foundation, RICS Foundation, London.

Syms, P. (2002) *Land, Development & Design*. Blackwell Publishing, Oxford.
Syms, P. & Knight, P. (2000) *Building Homes on Used Land*. RICS Books, Coventry.
Syms, P. & Weber, B. (2003) *International approaches to the valuation of land and property affected by contamination*. RICS Foundation, London.
Tewdwr-Jones, M. (ed.) (1996) *British Planning Policy in Transition*. UCL Press, London.
Tewdwr-Jones, M. (2002) *The Planning Polity: Planning, Government and the Policy Process*. Routledge, London.
Thomas, H. (2000) *Race and Planning: the UK Experience*. Taylor and Francis, London.
Urban Task Force (1999) *Towards an Urban Renaissance*. E & FN Spon, London.
Wannop, U. (1995) *The Regional Imperative: Regional Planning and Governance in Britain, Europe and the United States*. Jessica Kingsley and the Regional Studies Association, London.
Weber, B. (1996) Stigma – Unquantified Risk? Paper presented at The Cutting Edge 1996, a research conference of the Royal Institution of Chartered Surveyors. The University of the West of England, Bristol, September.
Weber, B. (1997) The Valuation of Contaminated Land. *Journal of Real Estate Research*, vol. 14, no. 3, pp. 379–398.
Weber, B. (1998) Stigma: Quantifying Murphy's Law. *Urban Land*, June, p. 12.
Whitney, D. (2003) Development Plan-making and Public Engagement: Responding to and Managing Public Expectations. Paper to Yorkshire Conference Series event on 'Plan Making and Review' at Sheffield, UK, 19 November.
Williams, R. H. (1996) *European Union Spatial Policy and Planning*. Paul Chapman, London.
Wilson, A. R. (1992) *Environmentally Impaired Valuation: A Team Approach to a Balance Sheet Presentation*. Technical Report: Measuring the Effects of Hazardous Materials Contamination on Real Estate Values: Techniques and Applications. Appraisal Institute, pp. 23–42.
Wilson, A. R. (1994) The Environmental Opinion: basis for an impaired value opinion. *Appraisal Journal*, July, pp. 410–423.
Wilson, A. R. (1996) Emerging Approaches to Impaired Property Valuation. *Appraisal Journal*, April, pp. 155–170.
Wiltshaw, D. G. (1998) Stigma, perception and the remediation of contaminated land. *Journal of Property Research*, vol. 15, no. 4, pp. 285–303.
Wright, M. (2003) 'Norwich throws a lifeline to homes without cover: a new mapping technique assesses risk house-by-house', in *The Daily Telegraph*, Money section, 29 November.
Wyatt, P. & Ralphs, M. (2003) *GIS in Land and Property Management*. Spon Press, London.

Department of the Environment – Industry Profiles

The DOE Industry Profiles provide developers, local authorities and anyone else interested in contaminated land with information on the processes, materials and wastes associated with individual industries. They also provide information on the contamination which might be associated with specific industries, factors that affect the likely presence of contamination, the effect of mobility of contaminants and guidance on potential contaminants. They are not definitive studies but they introduce some of the technical considerations that need to be borne in mind at the start of an investigation for possible contamination.

Airports (ISBN 1 85112289 3)
Animal and animal products processing works (ISBN 1 85112238 9)
Asbestos manufacturing works (ISBN 1 85112231 1)
Ceramics, cement and asphalt manufacturing works (ISBN 1 85112290 7)
Chemical works: coatings (paints and printing inks) manufacturing works (ISBN 1 85112291 5)
Chemical works: cosmetics and toiletries manufacturing works (ISBN 1 85112292 3)
Chemical works: disinfectants manufacturing works (ISBN 1 85112293 1)
Chemical works: explosives, propellants and pyrotechnics manufacturing works (ISBN 1 85112237 0)
Chemical works: fertiliser manufacturing works (ISBN 1 85112289 3)
Chemical works: fine chemicals manufacturing works (ISBN 1 85112235 4)
Chemical works: inorganic chemicals manufacturing works (ISBN 1 85112295 8)
Chemical works: linoleum, vinyl and bitumen-based floor covering manufacturing works (ISBN 1 85112296 6)
Chemical works: mastics, sealants, adhesives and roofing felt manufacturing works (ISBN 1 85112296 6)
Chemical works: organic chemicals manufacturing works (ISBN 1 85112275 3)
Chemical works: pesticides manufacturing works (ISBN 1 85112274 5)
Chemical works: pharmaceuticals manufacturing works (ISBN 1 85112236 2)

Chemical works: rubber processing works (including works manufacturing tyres or other rubber products) (ISBN 1 85112234 6)
Chemical works: soap and detergent manufacturing works (ISBN 1 85112276 1)
Dockyards and dockland (ISBN 1 85112298 2)
Engineering works: aircraft manufacturing works (ISBN 1 85112299 0)
Engineering works: electrical and electronic equipment manufacturing works (including works manufacturing equipment containing PCBs) (ISBN 1 85112300 8)
Engineering works: mechanical engineering and ordnance works (ISBN 1 85112233 8)
Engineering works: railway engineering works (ISBN 1 85112254 0)
Engineering works: shipbuilding, repair and shipbreaking (including naval shipyards) (ISBN 1 85112277 X)
Engineering works: vehicle manufacturing works (ISBN 1 85112301 6)
Gasworks, coke works and other coal carbonisation plants (ISBN 1 85112232 X)
Metal manufacturing, refining and finishing works: electroplating and other metal finishing works (ISBN 1 85112278 8)
Metal manufacturing, refining and finishing works: iron and steelworks (ISBN 1 85112280 X)
Metal manufacturing, refining and finishing works: lead works (ISBN 1 85112230 3)
Metal manufacturing, refining and finishing works: non-ferrous metal works (excluding lead works) (ISBN 1 85112302 4)
Metal manufacturing, refining and finishing works: precious metal recovery works (ISBN 1 85112279 6)
Oil refineries and bulk storage of crude oil and petroleum products (ISBN 1 85112303 2)
Power stations (excluding nuclear power stations) (ISBN 1 85112281 8)
Pulp and paper manufacturing works (ISBN 1 85112304 0)
Railway land (ISBN 1 85112253 2)
Road vehicle fuelling, service and repair: garages and filling stations (ISBN 1 85112305 9)
Road vehicle fuelling, service and repair: transport and haulage centres (ISBN 1 85112306 7)
Sewage works and sewage farms (ISBN 1 85112282 6)
Textile works and dye works (ISBN 1 85112307 5)
Timber products manufacturing works (ISBN 1 85112308 3)
Timber treatment works (ISBN 1 85112283 4)
Waste recycling, treatment and disposal sites: drum and tank cleaning and recycling plants (ISBN 1 85112309 1)
Waste recycling, treatment and disposal sites: hazardous waste treatment plants (ISBN 1 85112310 5)
Waste recycling, treatment and disposal sites: landfills and other waste treatment or waste disposal sites (ISBN 1 85112311 3)

Waste recycling, treatment and disposal sites: metal recycling sites (ISBN 1 85112229X)
Waste recycling, treatment and disposal sites: solvent recovery works (ISBN 1 85112312 1)
Profile of miscellaneous industries, incorporating:
 Charcoal works, Dry-cleaners
 Fibreglass and fibreglass resins manufacturing works
 Glass manufacturing works
 Photographic processing industry
 Printing and bookbinding works (ISBN 1 85112313 X)

The profiles are priced at £10 each, although many are out of print, and are available from Defra Publications, c/o IFORCE Ltd, Imber Court Business Park, Orchard Lane, East Molesey, Surrey, KT8 OBZ (tel 08459 556000; fax 0208 957 5012; email defra@iforcegroup.com)

Useful Internet Addresses

Name	Internet address
Advantage West Midlands	www.advantagewm.co.uk
Appraisal Institute	www.appraisalinstitute.org
Association of Geo-technical and Geo-environmental Specialists (AGS)	www.ags.org.uk
British Geological Survey	www.bgs.ac.uk
British Property Federation	www.bpf.propertymall.com
Campaign to Protect Rural England	www.cpre.org.uk
Certa Insurance	www.certa.com
Chartered Institution of Environmental Health (CIEH)	www.cieh.org
Commission for Architecture in the Built Environment (CABE)	www.cabe.org.uk
Contaminated Land: Applications in the Real Environment (CL:AIRE)	www.claire.co.uk
Department for Environment, Food and Rural Affairs (defra)	www.defra.gov.uk
East Midlands Development Agency (EMDA)	www.emda.org.uk
East of England Development Agency (EEDA)	www.eeda.org.uk

English Partnerships	www.englishpartnerships.co.uk
Environment Agency	www.environment-agency.gov.uk
Environmental Data Services (ENDS)	www.ends.co.uk
Environmental Regulations (England and Wales)	www.netregs.environment-agency.gov.uk
Health and Safety Executive	www.hse.gov.uk
Inland Revenue	www.inlandrevenue.gov.uk
Institute of Environmental Management and Assessment (IEMA)	www.iema.net
Institution of Civil Engineers	www.ice.org.uk
Invest Northern Ireland	www.investni.com
Joseph Rowntree Foundation	www.jrf.org.uk
Landmark Information Group	www.landmarkinfo.co.uk
Law Society	www.lawsociety.org.uk
London Development Agency	www.lda.gov.uk
National House-Building Council (NHBC)	www.nhbc.co.uk
National Land Use Database (NLUD)	www.nlud.org.uk
North West Development Agency (NWDA)	www.nwda.co.uk
Office of the Deputy Prime Minister (ODPM)	www.odpm.gov.uk
One NorthEast	www.onenortheast.co.uk
Ordnance Survey	www.ordnancesurvey.co.uk
RICS Foundation	www.rics-foundation.org
Royal Institute of British Architects (RIBA)	www.riba.org
Royal Institution of Chartered Surveyors (RICS)	www.rics.org
Royal Town Planning Institute (RTPI)	www.rtpi.org.uk

Scottish Enterprise	www.scottish-enterprise.com
Scottish Environment Protection Agency	www.sepa.org.uk
South East England Development Agency (SEEDA)	www.seeda.co.uk
South West Regional Development Agency (SWRDA)	www.southwestrda.org.uk
Specialists in Land Contamination (SiLC)	www.silc.org.uk
Welsh Development Agency (WDA)	www.wda.co.uk
Yorkshire Forward	www.yorkshire-forward.com

Index